THE UNSOLVED UNIVERSE: CHALLENGES FOR THE FUTURE

T0224048

The Unsolved Universe:
Challenges for the Future

JENAM 2002

Edited by

Mário J.P.F.G. Monteiro

Centro de Astrofísica da Universidade do Porto,
Portugal

KLUWER ACADEMIC PUBLISHERS
DORDRECHT / BOSTON / LONDON

A C.I.P. Catalogue record for this book is available from the Library of Congress.

ISBN   978-90-481-6447-9

Published by Kluwer Academic Publishers,
P.O. Box 17, 3300 AA Dordrecht, The Netherlands.

Sold and distributed in North, Central and South America
by Kluwer Academic Publishers,
101 Philip Drive, Norwell, MA 02061, U.S.A.

In all other countries, sold and distributed
by Kluwer Academic Publishers,
P.O. Box 322, 3300 AH Dordrecht, The Netherlands.

Printed on acid-free paper

Table of Contents

vi

4. Challenges for the Future

Foreword

The *Joint European and National Astronomical Meeting* (JENAM) of 2002, was held in Porto - Portugal (2-7 September 2002), corresponding to the *11th Meeting of the European Astronomical Society* (EAS) and the *12º Encontro Nacional de Astronomia e Astrofísica* (12ENAA) of the *Sociedade Portuguesa de Astronomia* (SPA).

Portugal has a small and young community of researchers in Astronomy. This meeting have had an important role in marking the beginning of what we expect to be a new phase for Astronomy in Portugal. The fact that we have chosen to address "'the future" reflects this will of the Portuguese community to share and discuss our commitment for the next decades with our colleagues.

The meeting, titled "**The Unsolved Universe: Challenges for the Future**", aimed at discussing some of the major research programmes and objectives for the next decades. The scientific programme included the plenary sessions (invited reviews and highlight talks), whose contributions are published in this book, and several workshops on more specific topics. The workshops organised within JENAM 2002 were:

- *The Very Large Telescope Interferometer: Challenges for the Future*

- *GAIA and DIVA Photometry: Towards the Fine Structure of the HR Diagram?*

- *JETS 2002: Theory and Observations in YSOs*

- *From Observations to Self-Consistent Modelling of the ISM in Galaxies*

- *Galactic Dynamics*

- *Galaxy Evolution in Groups and Clusters*

- *3K, SN's, Clusters: Hunting the Cosmological Parameters with Precision Cosmology*

- *The Cosmology of Extra Dimensions and Varying Fundamental Constants*

The meeting was very successful in stimulating the interaction between different fields of astronomy, in the same event, by bringing together researchers from usually separated communities. By using this well establish annual event for European astronomers to meet on a large scale, we have also aimed at providing a privileged setup for young researchers to discuss the forthcoming major European projects in Astronomy. The numbers of registered young researchers confirms that we have been successful in this goal.

This book reflects the multi-disciplinary and interaction that took place in the meeting. The reviews and highlights published here cover some of the major fields and projects which will determine the research in astronomy in the next decades. The highlights have been presented by young astronomers from several European countries, selected from a list of proposals submitted by different European institutions.

Many people and organisations contributed towards making JENAM 2002 a great success. We gratefully acknowledge the invaluable and tireless work of the Local Organising Committee (as someone has said "it was the dream team"), and the Conveners of the individual workshops for the excellent collaboration towards making possible the organisation of the JENAM. The meeting as a whole, and the workshops in particular, have had the financial support from several sponsors (see list below) and the contribution of several institutions and individuals which have made possible this meeting to happen.

Of course the whole success of the meeting was only possible thanks to the participants themselves and in particular the review speakers and those who presented oral and poster contributions. In all, we have had about 430 participants, whose more than 280 oral contributions and 100 poster contributions have been published producing seven thematic books of proceedings. Thank you all!

Mário J.P.F.G. Monteiro
Porto, 2003 July 8.

Committees and Sponsors

Scientific Organising Committee:

Domingos Barbosa	*Portugal*
Harvey Butcher (Co-Chairman, EAS)	*The Netherlands*
Anatol M. Cherepashchuk	*Russia*
Cesare Chiosi	*Italy*
Michel Dennefeld	*France*
João Fernandes	*Portugal*
Joachim Krautter	*Germany*
João J.G. Lima	*Portugal*
Mário J.P.F.G. Monteiro (Co-Chairman, SPA)	*Portugal*
Margarida Serote Roos	*Portugal*

Local Organising Committee:

Vitor M. Costa	*CAUP & ISEP, Porto*
João Fernandes	*OAUC, Coimbra*
Jorge Filipe Gameiro	*CAUP & DMA-FCUP, Porto*
João J.G. Lima	*CAUP & DMA-FCUP, Porto*
Catarina Lobo	*CAUP & DMA-FCUP, Porto*
Carlos J.A.P. Martins	*CAUP, Porto & Cambridge, U.K.*
Mário J.P.F.G. Monteiro (Chairman)	*CAUP & DMA-FCUP, Porto*
Margarida Serote Roos	*CAAUL, Lisbon*

Organisation:

European Astronomical Society (*www2.iap.fr/eas*)
Sociedade Portuguesa de Astronomia (*www.sp-astronomia.pt*)

Host:

Centro de Astrofísica da Universidade do Porto (*www.astro.up.pt*)

Sponsors (Plenary Sessions and Workshops):

Centro de Astrofísica da Universidade do Porto
(*www.astro.up.pt*)
Centro de Astronomia e Astrofísica da Universidade de Lisboa
(*www.oal.ul.pt/caaul*)
Centro Multidisciplinar de Astrofísica do Instituto Superior Técnico
(*centra.ist.utl.pt*)
European Southern Observatory (*www.eso.org*)
European Space Agency (*www.esa.int*)
Fundação para a Ciência e a Tecnologia (*www.fct.mces.pt*)
    Ministério da Ciência e do Ensino Superior
Observatório Astronómico da Universidade de Coimbra
(*www.astro.mat.uc.pt/obsv/*)
Universidade do Porto - Reitoria (*www.up.pt*)

Venues were provided by:

Centro de Astrofísica da Universidade do Porto (*www.astro.up.pt*)
Faculdade de Arquitectura da Universidade do Porto (*www.arq.up.pt*)
Faculdade de Ciências da Universidade do Porto (*www.fc.up.pt*)
    Departamento de Física (*www.fc.up.pt/fis*)
    Departamento de Matemática Aplicada (*www.ma.fc.up.pt*)
    Departamento de Matemática Pura (*www.fc.up.pt/mp*)

List of participants

ACKE, BRAM ————————————————————— Belgium
AFONSO, JOSÉ M. ————————————————— Portugal
AGUIAR, PAULO J.A.C. ————————————— Portugal
ALLEN, STEVEN W. ———————————————— U.K.
ALVES, JOÃO F. ————————————————— Germany
AMARO-SEOANE, PAU ———————————— Germany
ANDREON, STEFANO ———————————————— Italy
ANGELA, JOSÉ A.C. ————————————————— Portugal
ANTÓN, SÓNIA ———————————————————— Portugal
ARDI, ELIANI ————————————————————— Germany
AUGUSTO, PEDRO M.E.R.S. ——————————— Portugal
AVELINO, PEDRO P. ——————————————— Portugal

BAAN, WILLEM A. ———————————— The Netherlands
BACCIOTTI, FRANCESCA ——————————————— Italy
BACELLS, MARC ——————————————————— Spain
BAGLIN, ANNIE ———————————————————— France
BALAZS, LAJOS G. —————————————————— Hungary
BALLESTEROS, JAVIER ———————————————— Mexico
BARABASH, ALEXANDER S. ————————————— Russia
BARBERA, CARLOS ——————————————————— Spain
BARBOSA, DOMINGOS ——————————————— Portugal
BARROW, JOHN D. —————————————————— U.K.
BARTLETT, JAMES G. ———————————————— France
BASSETT, BRUCE A. ———————————————— U.K.
BAUGH, CARLTON ——————————————————— U.K.
BAUMGARDT, HOLGER ———————————————— Japan
BEÇA, LUÍS M.G. ————————————————— Portugal
BECK, RAINER ———————————————————— Germany
BELIKOV, ANDREI ———————————————— Germany
BERKHUIJSEN, ELLY M. ——————————————— Germany

BEST, PHILIP N.	U.K.
BHATTACHARYA, NANDINI	The Netherlands
BIESIADA, MAREK	Poland
BIRK, GUIDO T.	Germany
BIVIANO, ANDREA	Italy
BLANCHARD, ALAIN	France
BLINNIKOV, SERGEY	Russia
BODEN, ANDREW F.	U.S.A.
BOILY, CHRISTIAN M.	France
BOOMSMA, RENSE	The Netherlands
BOONE, FRÉDÉRIC	France
BORGANI, STEFANO	Italy
BOUCHET, FRANÇOIS R.	France
BRAVO ALFARO, HECTOR	Mexico
BREGMAN, JOEL	U.S.A.
BREITSCHWERDT, DIETER	Germany
BRINCHMANN, JARLE	Germany
BROUGH, SARAH	U.K.
BRUELL, MARTIN	Germany
BUCHER, MARTIN	U.K.
BUSSÓNS GORDO, JAVIER	France
BUTCHER, HARVEY	The Netherlands
CABRIT, SYLVIE	France
CACCIARI, CARLA	Italy
CAMENZIND, MAX	Germany
CANAVEZES, ALEXANDRE G.D.R.S.	U.K.
CARDWELL, ANDREW	Spain
CARREIRA, MARCO A.M.	Portugal
CARTIER, CYRIL	Switzerland
CARVALHO, CARLA	U.K.
CEMELJIC, MILJENKO	Germany
CHELLI, ALAIN	France
CHEREPASHCHUK, ANATOL M.	Russia
CHRISTLEIN, DANIEL	U.S.A.
CHRYSOSTOMOU, ANTONIO	U.K.
CIMATTI, ANDREA	Italy

FREYBERG, MICHAEL J. _____ *Germany*

FRIDLUND, MALCOLM _____ *The Netherlands*

FRITZ, ALEXANDER _____ *Germany*

FROEBRICH, DIRK _____ *Germany*

FUJII, YASUNORI _____ *Japan*

GAL'TSOV, DMITRI V. _____ *Russia*

GAMEIRO, JORGE FILIPE S. _____ *Portugal*

GARCIA, PAULO J.V. _____ *Portugal*

GARCIA-GOMEZ, CARLOS _____ *Spain*

GARDINER, THOMAS A. _____ *U.S.A.*

GAZOL, ADRIANA _____ *Mexico*

GERMANI, CRISTIANO _____ *U.K.*

GHESQUIERE, CLAUDE _____ *France*

GIANNINI, TERESA _____ *Italy*

GIARD, MARTIN _____ *France*

GIL, CARLA S.C. _____ *Portugal*

GIRARDI, MARISA _____ *Italy*

GLINDMANN, ANDREAS _____ *Germany*

GODDI, CIRIACO _____ *Italy*

GOMEZ DE CASTRO, ANA I. _____ *Spain*

GOMEZ, GILBERTO C. _____ *U.S.A.*

GONÇALVES, DENISE R. _____ *Spain*

GONZÁLEZ-DÍAZ, PEDRO F. _____ *Spain*

GONZALEZ, ANTHONY H. _____ *U.S.A.*

GREGG, MICHAEL D. _____ *U.S.A.*

GURVITS, LEONID _____ *The Netherlands*

GUSEV, ALEXANDER _____ *Russia*

GUTIÉRREZ, CARLOS _____ *Spain*

HAGIWARA, YOSHIAKI _____ *The Netherlands*

HAMIDOUCHE, MOURAD _____ *France*

HANASZ, MICHAL _____ *Poland*

HANIFF, CHRISTOPHER A. _____ *U.K.*

HARFST, STEFAN _____ *Germany*

HARTIGAN, PAT _____ *U.S.A.*

HEBRARD, GUILLAURE _____ *France*

HEMPEL, ANGELA _____ *Germany*

HENNING, THOMAS ———————————————— *Germany*
HERBST, TOM ———————————————— *Germany*
HERDEIRO, CARLOS A.R. ———————————————— *Portugal*
HERVIK, SIGBJORN ———————————————— *U.K.*
HOCKSELL FIUZA, DIANA A. ———————————————— *Finland*
HOUZIAUX, LÉO N. ———————————————— *Belgium*
HOVHANNISYAN, MARTIK A. ———————————————— *Armenia*
HRON, JOSEF ———————————————— *Austria*

IMAI, HIROSHI ———————————————— *The Netherlands*
IMMELI, ANDREAS ———————————————— *Switzerland*
INFANTE, LEOPOLDO ———————————————— *Chile*
INSKIP, KATHERINE J. ———————————————— *U.K.*
IOVINO, ANGELA ———————————————— *Italy*
IVANCHIK, ALEXANDRE ———————————————— *Russia*

JAANISTE, JAAK ———————————————— *Estonia*
JACHYM, PAVEL ———————————————— *Czech Republic*
JENKINS, EDWARD B. ———————————————— *U.S.A.*
JERSAK, JIRI ———————————————— *Germany*
JESSEIT, ROLAND ———————————————— *Germany*
JORDI, CARME ———————————————— *Spain*
JUNGWIERT, BRUNO ———————————————— *Czech Republic*
JUST, ANDREAS S. ———————————————— *Germany*

KALBERLA, PETER M.W. ———————————————— *Germany*
KALVOURIDIS, TILEMAHOS ———————————————— *Greece*
KANNO, SUGUMI ———————————————— *Japan*
KATGERT, PETER ———————————————— *The Netherlands*
KAY, SCOTT T. ———————————————— *U.K.*
KEEGAN, RONAN M. ———————————————— *Ireland*
KENDALL, TIM R. ———————————————— *Portugal*
KHANZADYAN, TIGRAN ———————————————— *U.K.*
KHOCHFAR, SADEGH ———————————————— *Germany*
KITIASHVILI, IRINA ———————————————— *Russia*
KLEIN, ULRICH ———————————————— *Germany*
KLÖCKNER, HANS-RAINER ———————————————— *The Netherlands*
KNUDE, JENS K. ———————————————— *Denmark*

PERRIN, GUY S.G. _____ *France*

PERRYMAN, MICHAEL A.C. _____ *The Netherlands*

PESENTI, NICOLAS _____ *France*

PETROV, ROMAIN G. _____ *France*

PHELPS, RANDY L. _____ *U.S.A.*

PINHEIRO, FERNANDO J.G. _____ *Portugal*

PINTO, PAULO M.S.P. _____ *Portugal*

PLANA, HENRI M. _____ *Brazil*

POGGIANTI, BIANCA MARIA _____ *Italy*

POMPEI, EMANUELA _____ *Chile*

POPOVIC, LUKA C. _____ *Servia*

PREIBISCH, THOMAS _____ *Germany*

PRETO, MIGUEL _____ *U.S.A.*

PRINSINZANO, LOREDANA _____ *Italy*

PRZYGODDA, FRANK _____ *Germany*

PYO, TAE-SOO _____ *U.S.A.*

QUIRRENBACH, ANDREAS _____ *U.S.A.*

REGEV, ODED _____ *Israel*

REIS, RICARDO S.S.C. _____ *Portugal*

REVAZ, YVES _____ *Switzerland*

RIAZUELO, ALAIN _____ *France*

RIBEIRO, ANDRÉ M.S. _____ *Portugal*

RICHICHI, ANDREA _____ *Germany*

ROCHA, GRAÇA _____ *U.K.*

RODRIGUEZ ESPINOSA, JOSE M. _____ *Spain*

ROESER, SIEGFRIED _____ *Germany*

ROSATI, FRANCESCA _____ *Italy*

ROSEN, ALEX _____ *U.K.*

ROSOLOWSKY, ERIK _____ *U.S.A.*

RUIZ-LAPUENTE, PILAR _____ *Spain*

SABATINI, SABINA _____ *U.K.*

SALAS, LUIS _____ *Mexico*

SAMLAND, MARKUS _____ *Switzerland*

SANCISI, RENZO _____ *Italy*

SANTOS, NUNO C. _____ *Portugal*

SASAKI, MANAMI ————————————————— *Germany*

SAUTY, CHRISTOPHE ———————————————— *France*

SCHADE, DAVID J. ————————————————— *Canada*

SCHAHMANECHE, KYAN ———————————————— *France*

SCHILBACH, ELENA ———————————————— *Germany*

SCHILIZZI, RICHARD T. ———————————— *The Netherlands*

SCHINDLER, SABINE ———————————————— *Austria*

SCHOELLER, MARKUS ———————————————— *Chile*

SEIICHI, KATO ————————————————— *Japan*

SEROTE ROOS, MARGARIDA ———————————— *Portugal*

SHANG, HSIEN ————————————————— *Taiwan*

SHAVER, PETER ————————————————— *Germany*

SHUKUROV, ANVAR ———————————————— *U.K.*

SIEBERT, ARNAUD ———————————————— *France*

SILVA, ANTÓNIO J.C. ———————————————— *France*

SILVA, CAROLINE S. ———————————————— *Portugal*

SMITH, KESTER W. ———————————————— *Switzerland*

SMITH, MICHAEL D. ———————————————— *U.K.*

SMOOT, GEORGE F. ———————————————— *U.S.A.*

SODA, JIRO ————————————————— *Japan*

SOLORZANO-INARREA, CARMEN ————————— *U.K.*

STANKE, THOMAS ———————————————— *Germany*

STAVINSCHI, MAGDA ———————————————— *Romania*

STAWARZ, TUKASZ ———————————————— *Poland*

STEE, PHILIPPE ————————————————— *France*

SWINGS, JEAN-PIERRE ———————————————— *Belgium*

TAKAMI, MICHIHIRO ———————————————— *U.K.*

TANAKA, YASUO ———————————————— *Germany*

TEIXEIRA DE ALMEIDA, MARIA LUÍ ————— *Portugal*

TEIXEIRA, PAULA S. ———————————————— *Portugal*

TEMPORIN, SONIA G. ———————————————— *Austria*

THIEBAULT, ERIC M.M. ———————————————— *France*

TORRENTE-LUJAN, EMILIO ————————————— *Spain*

TOVMASSIAN, HRANT M. ———————————————— *Mexico*

TREU, TOMMASO L. ———————————————— *U.S.A.*

TRINCHIERI, GINEVRA ———————————————— *Italy*

Asteroseismology and Planet Finding:
The Ultra High Precision Photometry European Road Map

Annie Baglin
LESIA, Observatoire de Paris, FRE CNRS 2461, France

2002 November 26

Abstract. Detection of many the eigenmodes of pulsations in stars of all types and of small planets orbiting around a large number of stars are foreseen in a very near future thanks to the efforts made by the European community, particularly in France. Indeed several such space missions as well as ground based facilities will be operating soon. Many other subjects in stellar physics and stellar environment will also benefit from these completely new observations.

Keywords: seismology, extrasolar planets, observations, space

1. Introduction

Stellar variability is very common and known since the early days of astronomy; it is one of the favourite activities of amateur astronomers. However, only large amplitude variability has been accessed and with a very poor time sampling. The improvement of instruments, and in particular their stability, as well as the ability to go to space, open a new domain of exploration in astrophysics.

In this paper, we recall the scientific interest of stellar microvariability (intrinsic or environmental) on time scales from seconds to months as a new field which can and will address many different domains in stellar physics and planetology. We briefly review the major ones : Asteroseismology, Detection of planet and other orbiting bodies, Stellar activity and many other stellar, galactic and extragalactic studies.

Then we discuss the performances of the two major techniques of observation: spectroscopy from the ground and photometry from space at the utmost level of accuracy.

We describe the space projects already in construction or planned for the near future, and focus on the European Ultra High Precision Photometry Road Map, which gives presently the leadership to Europe in this domain. Later prospects are also sketched.

As many reviews have already dealt with the different questions addressed here, I will extensively refer to them.

M.J.P.F.G. Monteiro (ed.), The Unsolved Universe: Challenges for the Future, 1-14.

2. Scientific rationale

Presently we are aware of two major phenomena producing stellar microvari-ability: intrinsic pulsations and periodic variations induced by planets orbiting around the star. This is why these two scientific subjects are linked through the method of observation, though the scientific fields are quite different. Some other domains may also be addressed.

2.1. SEISMOLOGY

One can talk about seismology when one is dealing with a large number of modes of pulsation in the same star. These multiperiodic oscillations associ-ated to the eigenmodes of the object are known and studied extensively in the Sun and in white dwarfs, and to a lesser extend in Delta scuti stars and a few solar analogs.

Seismology is the theoretical tool to interpret these oscillations in terms of internal structure. It is the only method to "see" inside the stars (except the difficult measurement of neutrino fluxes). Then seismic observations are badly needed to understand stellar evolution.

Stellar evolution governs the evolution of galaxies and of the Universe. It controls for instance its continuous chemical evolution, as the stellar cores are the sites of transformation of hydrogen into other elements up to heavy metals. Though some trends are now firmly established, the description of this evolution in all stages and for all types of stars is far from being understood.

Many reviews have presented in details the state of the art in stellar mod-elling, and described the major present uncertainties (see e.g. Lebreton, 2000), concerning essentially hydrodynamical processes as e.g. convective energy transport, mixing processes, angular momentum transport, different hydro-dynamical instabilities.

The Sun is the only star for which the seismic technique has been exten-sively and successfully applied. Thanks to helioseismology, many improve-ments have been made. But the situation of the Sun is very specific, and does not cover the domain of physical conditions that prevail in different stars. So helioseismology has to be generalised to a large variety of stars, covering the HR diagram, but also different rotation regimes and different chemical compositions.

The seismic properties are best described in the Fourier domain, with three major quantities for each eigenmode: the **frequency** which is in principle quite easily related to the static model, the **amplitude** which is linked to non linear hydrodynamics of the interaction between the mode and the dynamics of the medium, the **line width** which gives the life time of the mode and is related to its energy budget.

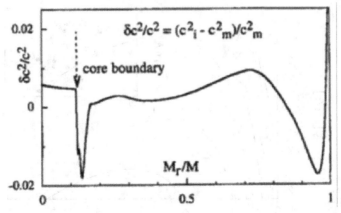

Figure 1. Simulation of the seismic analysis of a 1.45 M star, showing a very precise reconstruction of the sound velocity profile, allowing to fix the position of the core boundary from Roxburgh et al (2002).

The use of the seismic information to probe stellar interiors has been extensively documented already; see e.g. Weiss and Baglin (1993). The task is complex, as most of the stars have a more complicated structure than the Sun, leading to a less regular Fourier spectrum than in solar analogs as discussed e.g. in Thompson et al. (2002). In addition rotation (and eventually magnetic field) is not always sufficiently slow to be considered as a linear perturbator.

Simulations (see e.g. Roxburgh et al., 2002, Goupil et al., 2002) have proven the efficiency of the method in the stellar case as illustrated in Figure 1, if the Fourier spectrum is well understood and the quantum numbers of the eigenmodes are unambiguously identified. However, there is still a lot of work to be done to interpret seismic data of many different types of stars as expected from the future projects, as COROT and EDDINGTON (see 3.3).

Amplitudes are indicators of the energetic content of the turbulence exciting the modes, and as shown on Figure 2, they can be used to built diagnostics of the type of turbulence at work.

2.2. EXTRASOLAR PLANET FINDING

Since the discovery of the first extrasolar planet in 1995 by Mayor and Queloz (1995), more than 100 planets have been discovered (see e.g. Extrasolar planet Encyclopedia, maintained by J. Schneider at http://www.obspm.fr/encycl) .

But these planets are all giants, as seen on Figure 3 due to the present threshold of detection and close to their parent star. The present state of the art of the theory cannot explain these facts. The migration phenomenon is not yet fully satisfactory and depends upon many unknown physical parameters.

To built a reasonable theory of planetary systems formation, which can produce predictions about the population of planets and in particular the

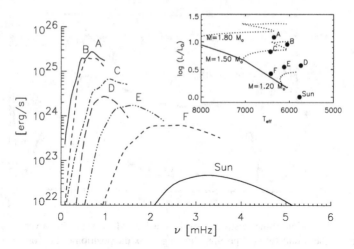

Figure 2. Amplitudes of solar like oscillations along the HR diagram, and dependence of the amplitudes to the turbulent energy from Samadi et al. (2002)

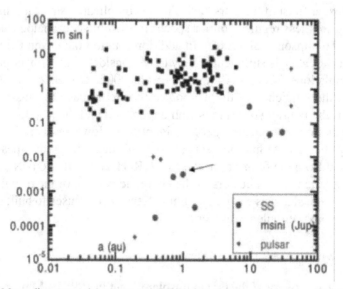

Figure 3. Mass, distance to the parent star distribution of presently known extrasolar planets (squares); the planets of the solar system are indicated by dots.

smaller ones (see e.g. Léger et al., 2002), new programmes of observations are needed, aiming at detecting planets in a very wide range of masses, distances and stellar parameters of the parent star.

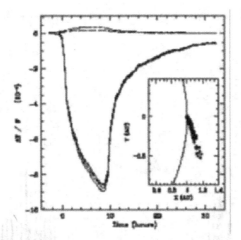

Figure 4. Photometric perturbations due to the infall of a comet from a stellar disk, from Lecavelier et al. (1999).

2.3. STELLAR ACTIVITY AND OTHER TOPICS

Stellar microvariability on time scales from seconds to months addresses many other stellar topics. One of the major ones is stellar activity. As shown by the solar space experiments like VIRGO/SoHO, luminosity variations somehow erratic exist in this domain of frequencies.

But, the associated physical process is still difficult to identify (Carpano et al., 2002). The observation in other stars of different external structures and outer convective zones will give clues to this question.

Many other fields will benefit from these variability studies, as listed for instance in (Gilmore, 2002). Let us cite solar system minor bodies, Kuiper belt objects, rotational modulation in young stars, infall of comets from stellar discs (as illustrated on Figure 4), stellar disks and pre-planetary systems, accretion disks stellar systems, X-rays pulsar systems, eclipsing binaries, Seyfert galaxies, AGN, Supernovae and γ ray bursts, QSO and galaxy counts, low surface brightness galaxies, gravitational lensing......

3. Methods of observation

The tiny variations of the stellar properties produced by intrinsic microvariability, or by an orbiting planet, are detectable more easily in the optical domain, where most of the stellar energy is concentrated. As relative time variations are looked for, long uninterrupted runs are necessary in all cases.

Two major techniques are available at present to observe these variations at the required level, and used for both objectives.

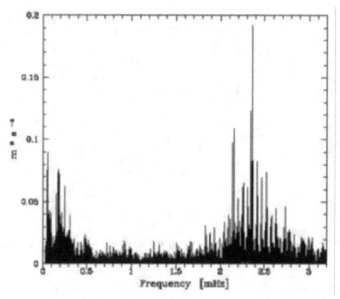

Figure 5. Power spectrum of radial velocity variations in α *Cen* from Bouchy et al. (2002), showing an important excess of power between 1.5 and 3 mHz, and well defined spectral features.

3.1. HIGH PRECISION VELOCIMETRY

Using a high precision and very stable spectrograph, this technique measures the variations of the stellar radial velocity.

To detect stellar oscillations analogous to those observed in the Sun an accuracy of a few cm/s is required in a few days. To access planets the level of a few m/s over one exposure is necessary.

The major difficulties lie on the size of the telescope (several metres diameter) to reach a sufficient number of stars, and on its accessibility over long uninterrupted periods; the need for a better coverage than provided by a unique site favours networks coordinated in longitude. A very high instrumental stability over long periods has to be preserved.

Oscillations have been detected by this method on a few bright stars, the prototype being α *Cen* (Bouchy et al., 2002), as shown on Figure 5. Several tens of modes are detected; the simple internal structure of this solar-like star allows mode identification and some preliminary seismic interpretations.

All planets known at present, as seen on Figure 3, have been observed with this technique on 2 m class telescopes, essentially by two independent groups: Mayor and collaborators, and the Marcy and Butler team. The progress foreseen with the new generation spectrographs on 4 m class telescopes, like HARPS, will allow a $1m/s$ precision or even better....

3.2. ULTRA HIGH PRECISION RELATIVE PHOTOMETRY

In this technique, there exists an unavoidable limit to the signal to noise ratio S/N, imposed by the photon noise. Generally, projects propose to reach a S/N value as close as possible to this limit, at least of the same order of magnitude. As this S/N is proportional to the square root of the counted photons in a given exposure, strong efforts are focussed on collecting as many photons as possible. In particular, one always uses a wavelength interval as large as possible, limited by the optical parts in the blue and by the detectors in the red.

3.2.1. *In seismology*

The amplitude of the solar modes are of a few ppm and their life time several days. So the noise level has to remain smaller than 1 ppm in a few days. This is not possible from the ground in this frequency domain due to atmospheric fluctuations.

The signal to noise ratio for a periodic signal is expressed in this case as:

$$S/N = \frac{nta^2}{4} \tag{1}$$

where a is the amplitude, n the mean counting rate (expressed in photons per s) and $t = inf(T,\tau)$ with T the total duration of the observation and τ the life-time of the mode.

Assuming that the limit of the photon noise is reached (which implies also strong constraints on the stability of the instrument), intermediate size telescope are sufficient, as seen on Figure 6. For instance a 30 cm telescope can detect solar like oscillations down to magnitude 7, which corresponds to a sample already sufficient to probe the main sequence and its surroundings, while a 90 cm aperture allows to go down to 12th magnitude, and then to access many different and exotic stars.

3.2.2. *In planet detection*

The brightness relative variation due to the transit of a planet in front of its parent star, if the line of sight is close to the plane of the orbit, is:

$$\Delta F/F = \left(\frac{R_{pl}}{R_{st}}\right)^2 \tag{2}$$

where R_{pl} and R_{st} are the radius of the planet and of the star. In the case of Jupiter and the Sun $\Delta F/F = 10^{-4}$, and for the Earth and the Sun only 8×10^{-5}. The transit lasts 31 and 14 hours respectively.

This means that telluric planets are detectable if the photon noise remains smaller than that during that period. This condition is fulfilled for a 14th magnitude star and a 25 cm telescope.

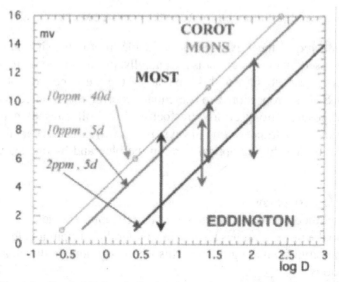

Figure 6. Detection threshold of periodic signals as a function of the collecting area and the magnitude of the star, for different values of the amplitude and of the modes life time.

Evidently the periodicity of the orbital motion helps in the detection process and favours planets close to their parent star.

3.3. COMPLEMENTARITY

For planet detection, the transit measures the radius of the planet, while the velocimetry measurements give the mass. This is nicely illustrated by the famous system HD 209456 which has been detected and observed in radial velocity from the ground, and then with the HST in photometry, as the orbital plane is close to the line of site (Figure 7). These two observations allow to determine the mean density of 0.31 $g\ cm^{-3}$, indicating the gaseous nature of the body.

In seismology the use of both techniques can be interesting too, as the observational bias are different, and lead to different sensitivities on the types of modes: while photometry is only sensitive to almost radial global modes, spectroscopy favours intermediate degrees, specially in rotating objects.

3.4. THE ULTRA HIGH PRECISION PHOTOMETRY ROAD MAP

While spectroscopy is definitely a ground based technique, photometry has to be performed from space, due to the atmospheric perturbations at that level of accuracy.

This technique is (was) not classical for space activities, devoted essentially to electromagnetic domains not accessible from the ground.

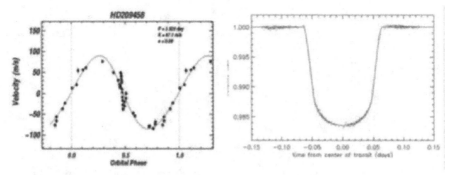

Figure 7. Radial velocity and photometric observations of HD 209456, from Naef et al.(2001) and Charbonneau et al. (2002).

Figure 8. EVRIS

3.4.1. *Some words of history*

The first idea to perform stellar relative photometry with a very high accuracy, e.g. from space goes back to the early 80's in France, as described in Roxburgh, 2002 . The first project to be accepted, funded and built was EVRIS (Etude de la variabilité et de la Rotation des Intérieurs Stellaires) (see Baglin et al., 1993). It has been launched on October 1996 on board of the Russian spacecraft Mars96, but due to a failure of an engine, it crashed just after launch.

An improved version of this instrument called MOST will fly soon, built by the Canadian Space Agency (see e.g. http://www.astro.ubc.ca/MOST/).

In the meantime, a second generation mission has been proposed by the French team in the framework of the "Petites missions" programme initiated by CNES in 1993 (Catala et al., 1993). After many difficulties this project has finally been selected in 2000 as COROT, with a more ambitious programme, including for the first time both topics, asteroseismology and telluric planet finding.

At the european level at ESA many asteroseismology proposals have been presented as PRISMA (1993), STARS (1996) and finally EDDINGTON (2000), which deals also with planet finding.

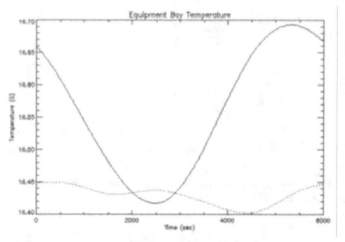

Figure 9. Variations of the temperature of the readout electronics boxes along one orbit, kept below 0.2 degrees, as specified, and measured continuously to allow efficient corrections.

In the framework of the danish satellite programme a microsatellite of seismology is under selection as RØMER/MONS, aiming at the observation of several bright solar-like stars (Kjeldsen et al., 2002).

3.4.2. COROT

This mission has been extensively described (see e.g. Baglin et al., 2002 and references therehein). The launch is planned at the end of 2005.

COROT belongs to the class of the minisatellites, with a total weight of 600 *kg*. The PROTEUS platform, used in this programme implies a low earth orbit. In this case long duration runs of half a year are possible, but only in limited regions of the sky. However, COROT will be able to access the seismic properties of stars covering a large part of the HR diagram, though focusing on the main sequence and its vicinity. For planet detection, it will observe in the same regions, several ten thousand dwarf faint stars, providing continuous light curves of very high photometric accuracy.

A very severe control of the instrument is necessary to be able to correct from the environmental perturbations (Auvergne et al., 2002), as illustrated on Figure 9.

The choice of the seismology targets is based on simulations using hare and hounds exercises (Figure 10), to qualify the best candidates, in the limits of all the constraints of the mission, including the exoplanet programme criterion.

3.4.3. EDDINGTON

The European project EDDINGTON, now in its final phase of selection, is more ambitious than COROT. The launch date is 2007/8. In seismology it

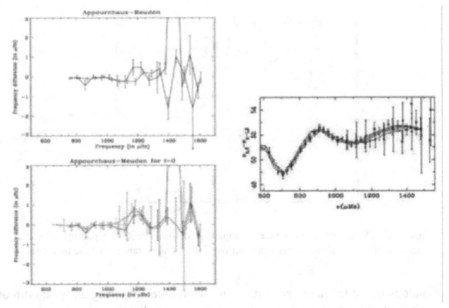

Figure 10. Preliminary results of Hare and Hounds exercises, showing the correct recovery of frequencies, and the ability to propose a stellar model close to the initial one, for a 1.4 solar mass main sequence model, from Berthomieu and Appourchaux (2002).

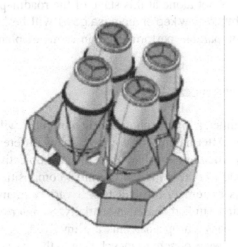

Figure 11. The most recent concept for the Eddington payload mounted on an Hershell bus, consisting of 4 identical coaligned 60 cm telescopes.

will perform an extended survey, being able to detect solar like oscillations down to magnitude 12. With its wide field of view, it will observe many clusters, and in each of them many different stars, allowing for differential seismology in objects of the same age and same initial chemical composition. In the planet finding mode, it will have the capability to detect earth

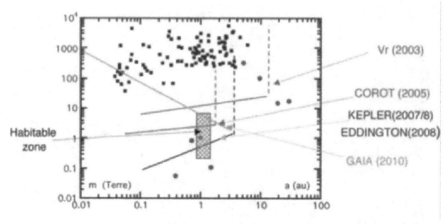

Figure 12. The planet hunting race. Comparison of the detection capability of extrasolar planets for the different missions as a function of mass and distance from the parent star.

analogs in the habitable zone down to magnitude 14 during a 3 years run on one field, close to the galactic plane. Uninterrupted photometric sequences of most of the stars of the field of view, down to magnitude 17 will be produced, addressing many topics in astrophysics.

But, Europe is not alone at this stage of the roadmap. The NASA project KEPLER (see http://www.kepler.arc.nasa.gov/) will be launched in the same period, with comparable performances in terms of planet detection (Figure 12).

3.5. FUTURE PROSPECTS

After the step aimed at by the photometric method, it will be necessary to use new techniques. After having "discovered" many different types of extrasolar planets and in particular earth-like ones, the next objective is to obtain images of the planet and to determine the chemical composition of its atmosphere. To perform this extremely difficult task a variety of instrumental concepts are already being studied, based essentially on interferometry, coronography, in which many european teams are involved. A very extended list is available at http://www.obspm.fr/encycl. Due to the extreme difficulty and important cost of such a project, it will probably be developed in a worldwide, may be "planetary" cooperative framework. In Europe, extensive studies have started already several years ago, leading to the DARWIN concept (see e.g. http://sci.esa.int/home/darwin/).

In the mean time in stellar physics, the major breakthrough will come from both new generation spectrographs and polarimeters being built on the largest telescopes, as well as the immense and unique survey mission GAIA, which, by providing fundamental parameters of most of the stars of the galaxy and beyond, will complement all the seismology data gathered during the next decade and allow a very precise understanding of stellar evolution , and then of the Universe; as said by A. Weiss (2002), *The better we want to know the Universe, the better we have to understand stars* !

4. Conclusion

The possibility to perform ultra high precision photometry from space, at the photon noise level will allow fundamental progresses in major field in astrophysics. Unfortunately, due to programmatics and also failures, this route has been very long to develop. Within a few years and definitely before the end of this decade, many space missions will be flown, complemented by new ground based instruments.

The different projects take into account all the possibilities of the technique and will produce the data necessary for understanding the basics of our theoretical understanding, in planetary formation, and in stellar hydrodynamics. Other fields will also benefit from these missions as e.g. stellar activity.

We are definitely at the dawn of a golden age for these domains, and a strong effort is being made to be prepared to interpret the data, both on the theoretical side and through ground based complementary observations.

In this new area, Europe has been leader since the early beginning, and remains so... but other countries are now proposing analogous projects to be flown approximately at the same period, contributing to these very exciting fields of research.

References

Auvergne, M., Boisnard, L., Buey, J-T., Esptein, G., Hustaix, H., Jouret-Perl, M., Levacher, P., Berrivin, S., Baglin, A.: 2002, *SPIE, Astronomical Telescopes and Instrumentation*, Waikoloa, August 22-28, 4859-47.

Baglin, A., Weiss, W., Bisnovatyi-Kogan, G.: 1993, *IAU colloqiuum* **137**, A.S.P. Conf. Ser. **40**, 758.

Baglin, A., Auvergne, M., Barge, P., Catala, C., Michel, E., Weiss, W., and the COROT Team: 2002, *ESA-SP* **465**, 17.

Berthomieu, G., Appourchaux, T., and the COROT seismology Team: 2003, in *Asteroseismology Across the HR Diagram*, Thompson M.J., Cunha M.S., Monteiro M.J.P.F.G. (eds), Kluwer Academic Publishers, 465.

Bouchy, F., Carrier, F.: 2002, *A&A* **390**, 205.

Catala, C., Auvergne, M., Baglin, A.: 1993, *GONG Meeting*, Los Angeles.

Carpano, S., Aigrain, S., Favata, F.: 2003, *A&A* **401**, 743.

Charbonneau, D., Brown, T., Noyes, R., Gilliland, R.L.: 2002, *ApJ* **568**, 377.

Favata, F., Roxburgh, I.W., Gimenez, A.: 2002, *Proceedings of the first Eddington Workshop: Stellar Structure and Planet Finding*, *ESA-SP* **465**.

Gilmore, G.: 2002, *ESA-SP* **465**, 177.
Goupil, M.-J., Dziembovsky, W., Goode, P.R., Michel, E.: 1996, *A&A* **305**, 487.
Kjedsen, H., Christensen-Dalsgaard, J.: 2002, *A.S.P. Conf. Ser.* **259**, 630.
Lebreton, Y.: 2000, *ARA&A* **38**, 35 .
Lecavelier des Etangs, A., Vidal-Madjar, A., Ferlet, R.: 1999, *A&A* **343**, 916.
Léger, A., Labèque, A., Ollivier, M., Sekulic, P.: 2002, *ESA-SP* **465**, 171.
Mayor, M., Queloz, D.: 1995, *Nat* **378**, 355.
Naef, D., Latham, D.W., Mayor, M., Mazeh, T.: 2001, *A&A* **375** , L27.
Roxburgh, I.W., Vorontsov, S.: 2002, *ESA-SP* **465**, 349.
Roxburgh, I.W., 2002 , *ESA-SP* **465**, 11.
Samadi, R.D., Houdek, G., Goupil, M-J., Lebreton, Y., Baglin, A.: 2002, *ESA-SP* **465**, 87.
Thompson, M., Christensen-Dalsgaard, J.: 2002, *ESA-SP* **465**, 95.
Weiss, A.: 2002, *ESA-SP* **465**, 57 .
Weiss., W., Baglin, A.: 1993, *Inside the Stars*, A.S.P. Conf. Ser. **40**.

Extra-Solar Planets: Clues to the Planetary Formation Mechanisms

Nuno C. Santos and Michel Mayor
Geneva Observatory, CH-1290 Sauverny, Switzerland

2002 November 14

Abstract. Radial Velocity surveys have revealed up to now about 100 extra-solar planets ($M \sin i < 10\,M_{Jup}$) and $\sim$10 multi-planetary systems. The discovered planets present a wide variety of orbital elements and masses, which are raising many problems and questions regarding the processes involved in their formation. But the analysis of the distributions of orbital elements, like the period and eccentricity distributions is already giving some constraints on the formation of the planetary systems. Furthermore, the study of the planet host stars has revealed the impressive role of the stellar metallicity on the giant planet formation. The chemical composition of the molecular cloud is probably the key parameter to form giant planets. In this article we will review the current status of the research on this subject.

Keywords: extra-solar planets, orbital parameters, metallicity

1. Introduction

It was not until in 1995, following the discovery of the planet orbiting the solar-type star 51 Peg (Mayor & Queloz, 1995), that the search for extra-solar planets had its first success[1]. Today, about 100 extra-solar planetary systems have been unveiled around stars other than our Sun[2]. These discoveries, that include $\sim$10 multi-planetary systems, have brought to light the existence of planets with a huge variety of characteristics, opening unexpected questions about the processes of giant planetary formation.

The diversity of the discovered extra-solar planets is well illustrated in Figure 1. Unexpectedly, they don't have much in common with the giant planets in our own Solar System. Contrarily to these latter, the "new" worlds present an enormous and unexpected variety of masses and orbital parameters (astronomers were basically expecting to find "jupiters" orbiting at $\sim$5 A.U. or more from their host stars in *quasi*-circular trajectories). In fact they were not even supposed to exist according to the traditional paradigm of giant planetary formation (Pollack et al., 2001). Their masses vary from sub-saturn to various times the mass of Jupiter. Some have orbits with semi-major axis smaller than the distance from Mercury to the Sun, and except for the closest

---

[1] Before this discovery, only planets around a pulsar had been detected (Wolszczan & Frail, 1992); these are probably second generation planets, however.

[2] See table at http://obswww.unige.ch/~naef/who_discovered_that_planet.html

M.J.P.F.G. Monteiro (ed.), The Unsolved Universe: Challenges for the Future, 15–24.
© 2003 *Kluwer Academic Publishers.*

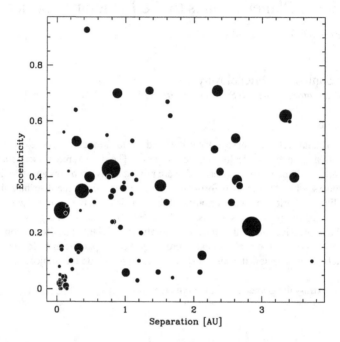

Figure 1. Eccentricity vs. orbital separation for the known planetary companions to solar-type stars. The size of the symbols is proportional to the minimum mass of the companion. This plot clearly shown the huge diversity of planetary companions found to date

companions, they generally follow eccentric trajectories, contrary to the case of the giant-planets in the Solar System.

But the relatively large number of discovered planets is already permitting us to undertake the first statistical studies of the properties of the exo-planets, as well of their host stars. This is bringing new constraints to the models of planet formation and evolution. In the rest of this article we will review the current results on the planetary searches, and in particular we will focus on the observational constraints the new discoveries are bringing.

2. Clues from the orbital parameters

2.1. PERIOD DISTRIBUTION

One of the most interesting problems that appeared after the first planets were discovered has to do with the proximity to their host stars. In contrast with the current observations, giant planets were previously thought to form (and be present) at distances of a few A.U. from their host stars (Pollack et al., 2001). In order to explain the newly found systems, several mechanisms have thus

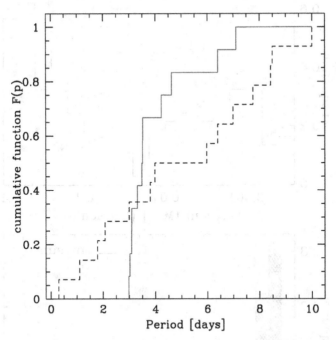

Figure 2. Cumulative distributions of periods smaller than 10 days for planetary (solid line) and stellar companions (dashed line) to solar type dwarfs

been proposed. Current results show that *in situ* formation is very unlikely (Bodenheimer et al., 2000), and we need to invoke inward migration, either due to gravitational interaction with the disk (Goldreich & Tremaine, 1980, Lin et al., 1996, Ward, 1997, Murray et al., 1998) or with other companions (Rasio & Ford, 1996, Lin & Ida, 1997) to explain the observed orbital periods.

Although still strongly biased for the long period systems, the period distribution of the extra-solar planetary companions can already tell us something about the planetary formation and evolution processes. This is particularly true for the short period systems, for which the biases are not so important. In particular, one of the most impressive features present in the current data is the clear pile-up of planetary companions with periods around 3 days, and the absence of any system with a period shorter than this (Figure 2). This result, that is in complete contrast with the period distribution for stellar companions, means that somehow the process involved in the planetary migration makes the planet "stop" at a distance corresponding to this orbital period. To explain this fact, several ideas have been presented, invoking e.g. a magnetospheric central cavity of the accretion disk, tidal interaction with the host star, Roche-lobe overflow by the young inflated giant planet, or evaporation.

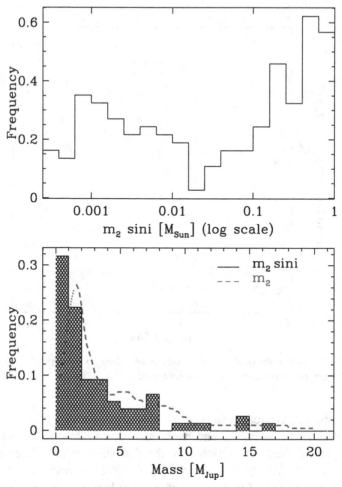

Figure 3. Mass function of companions to solar-type stars in *log* (top) and linear (bottom) scales. In the lower panel, the dashed line represents the result of a deconvolution of the observed distribution in order to take into account the effect of the orbital inclination. As in Jorissen et al. (2001)

2.2. THE MASS DISTRIBUTION

Another important clue concerning the nature of the now discovered planetary systems comes from their mass distribution. Although the radial-velocity technique is more sensitive to massive companions than to their lower mass counterparts, a look at the mass distribution (Figure 3) shows that this strongly rises towards the low mass regime – see Figure 3, lower panel.

Several conclusions may be taken from the plots. In the upper panel of Figure 3, the gap in the distribution, separating low mass stellar companions from the lower mass planets (often called the "brown dwarf desert") represents a strong evidence that these two populations are the result of different formation processes. Furthermore, we can see that the planetary mass distribution has a sharp cutoff for masses around $\sim 10\,M_{Jup}$ (Jorissen et al., 2001). This limit is clearly not related to the Deuterium-burning mass limit of $\sim 13\,M_{Jup}$, sometimes considered as the limiting mass for a planet[3]. As it was recently shown by Jorissen et al. (2001), this result is not an artifact of the fact that for most of the targets we only have minimum masses[4], but a real upper limit for the mass of the planetary companions discovered so far, since it is clearly visible in a deconvolved distribution (where the effect of the unknown orbital inclination was taken into account).

It is also very interesting to note that recent results strongly suggest that there is some relation between the mass of the companion and its orbital period: there seems to be a paucity of high-mass planetary companions (with $M > 2\,M_{Jup}$) orbiting in short period (lower than ~ 40-days) trajectories (Zucker & Mazeh, 2002, Udry et al., 2002). This trend, clearly significant, is nonetheless not found for those planets orbiting stars that have other stellar companions. These results are indeed compatible with the current ideas about planetary orbital migration (either due to an interaction with the disk or with other companions) – (Zucker & Mazeh, 2002). And although still not very constraining, the observed correlations will probably permit to help decide between the different models of planetary formation.

2.3. THE ECCENTRICITY DISTRIBUTION

One of the most enigmatic results to date is well illustrated in Figure 4. A first look at the figure shows that there are no clear differences between the eccentricity distributions of planetary and stellar binary systems. How then can this be fit into the "traditional" picture of a planet forming in a disk? For masses lower than $\sim 20\,M_{Jup}$, it has been shown that the interaction (and migration) of a companion within a gas disk has the effect of damping the eccentricity (Goldreich & Tremaine, 1980, Ward, 1997). This suggests that other processes, like the interaction between planets in a multiple system (Rasio & Ford, 1996) or between the planet and a disk of planetesimals (Murray et al., 1998), the simultaneous migration of various planets in a disk (Murray et al., 2002), or the influence of a distant stellar companion (Mazeh & Shaham, 1979, Cochran et al., 2000), may play an important role in defining the "final" orbital configuration. In this respect, one particularly interesting case of very high eccentricity (above 0.9) amongst the planetary companions is the planet around HD 80606 (Naef et al., 2001).

---

[3] This value is an arbitrary limit used as a possible "definition", but it is not related to the planetary formation physics.

[4] The unknown orbital inclination implies that we can only derive a minimum mass for the companion from the radial-velocity measurements.

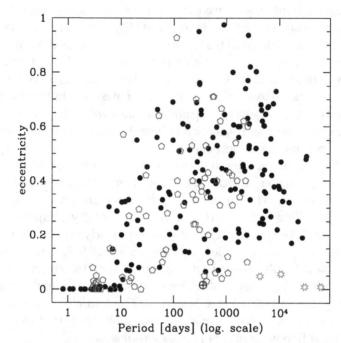

Figure 4. The $e - \log P$ diagram for planetary (open pentagons) and stellar companions (filled circles) to solar type field dwarfs. Starred symbols represent the giant planets of our Solar System, while the "earth" symbol represents our planet

Although still not clear, however, a close inspection of the Figure 4, permits to find a few differences between the eccentricities of the stellar and planetary companions. For example, for periods in the range of 10 to 30 days (clearly outside the circularization period by tidal interaction with the star), there are a few stars with planets having very low eccentricity, while no stellar binaries are present in this region. The same and even more strong trend is seen for longer periods, suggesting the presence of a group of planetary companions with orbital characteristics more similar to those of the planets in the Solar System. On the other hand, for the very short period systems, we can see some planetary companions with eccentricities higher than those found for stellar companions of similar period. These facts may be telling us that different formation and evolution processes took place: for example, the former group may be seen as a sign of formation (and evolution) in a disk, and the latter one as an evidence of the gravitational influence of a longer period companion on the eccentricity.

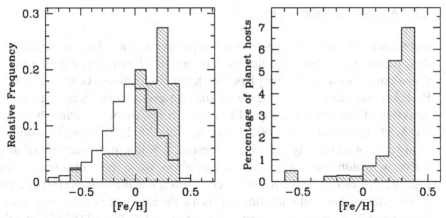

Figure 5. Left: metallicity distribution of stars with planets making part of the CORALIE planet search sample (shaded histogram) compared with the same distribution for the about 1000 non binary stars in the CORALIE volume-limited sample. *Right*: the result of correcting the planet hosts distribution to take into account the sampling effects. The vertical axis represents the percentage of planet hosts with respect to the total CORALIE sample. From Santos et al. (2002b).

3. The metal-rich nature of planet host stars

Up to now we have been reviewing the results and conclusions we have obtained directly from the study of the orbital properties and masses of the discovered planets. But another particular fact that is helping astronomers understand the mechanisms of planetary formation as to do with the planet host stars themselves. In fact, they were found to be particularly metal-rich, i.e. they have, in average, a metal content higher than the one found in stars without detected planetary companions (Gonzalez, 1998, Santos et al., 2001, Santos et al., 2002b). The most recent results (see Figure 5) seem to favour that this metallicity "excess" is original from the cloud that gave origin to the star/planetary system (Pinsonneault et al., 2001, Santos et al., 2001, Santos et al., 2002b).

A possible and likely interpretation of this may pass by saying that the higher the metallicity of the cloud that gives origin to the star/planetary system (and thus the dust content of the disk), the faster a planetesimal can grow, and the higher the probability that a giant planet is formed before the proto-planetary disk dissipates. In other words, the metallicity seems to be playing a key role in the formation of the currently discovered extra-solar planetary systems.

However, it is not known precisely how the influence of the metallicity is influencing the planetary formation and/or evolution; for example, the mass of the disks themselves, that can be crucial to determine the efficiency of planetary formation, is not known observationally with enough precision.

3.1. A CASE OF "POLLUTION"

But recent observations suggest also that planets might in fact be engulfed by their parent stars, whether as the result of orbital migration, or e.g. of gravitational interactions with other planet or stellar companions (Gonzalez, 1998). Probably the most clear evidence of such an event came recently from the detection of the lithium isotope ^{6}Li in the atmosphere of the planet-host star HD 82843 (Israelian et al., 2001, Israelian et al., 2002). This fragile isotope is easily destroyed (at only 1.6 million degrees, through (p,α) reactions) during the early evolutionary stages of star formation, when the proto-star is completely convective, and the relatively cool material at the surface is still deeply mixed with the hot stellar interior (this is not the case when the star reaches its "adulthood"). ^{6}Li is thus not supposed to exist in stars like HD 82843, and the simplest and most convincing way to explain its presence is to consider that planet(s), or at least planetary material, have fallen into HD 82843 sometime during its lifetime.

Recently, some authors have casted some doubts into the reality of the ^{6}Li detection in HD 82943 (Reddy et al., 2002). The most recent and detailed analysis seem, however, to clearly confirm the presence of this isotope (Israelian et al., 2002). The question is then turned to know whether this case is isolated or else if the fall of planetary material is a frequent outcome of the planetary formation process. The current results seem to suggest that at least the degree of stellar "pollution" is not incredibly high (Santos et al., 2001, Laws & Gonzalez, 2001, Santos et al., 2002c).

4. Conclusions

The study of extra-solar planetary systems is just giving its first steps. After only 7 years, we can say that at least 5% of the solar type dwarfs have giant planetary companions with masses as low as the mass of Saturn and orbital separations of a few Astronomical Units (the limits imposed by the current planetary search techniques). But the understanding of how giant planets are formed is still shaded in many points.

As we have seen above, the observed correlations between the orbital parameters of the newly found planets are giving astronomers a completely different view on the formation and evolution of the planetary systems. We no longer have the Sun as the only example, and today we have to deal with the peculiar characteristics of the "new" extra-solar planets: a huge variety of periods, eccentricities, masses.

Furthermore, the analysis of the chemical properties of the planet host stars is also giving us a lot of interesting information. These latter studies have revealed the crucial role the metallicity is playing into the formation of the currently found planetary systems, Furthermore, they are also showing us the wild side of planetary formation: planets might perish into their host stars.

As the planet search programs continue their way, many more planetary companions are expected to be discovered in the next few years. This will give us the opportunity to improve the statistical analysis, and to better understand the physics beyond the formation of the planetary systems.

Acknowledgements

We would like to thank the members of the Geneva extra-solar planet search group, D. Naef, F. Pepe, D. Queloz, S. Udry, as well as G. Israelian, R. Rebolo, and R.J. García López (from the IAC), who have largely contributed to the results presented here. We wish to thank the Swiss National Science Foundation (Swiss NSF) for the continuous support to this project. Support from Fundação para a Ciência e Tecnologia, Portugal, to N.C.S. in the form of a scholarship is gratefully acknowledged.

References

Bodenheimer, P., Hubickyj, O., Lissauer, J.J.: 2000, *Icarus* **143**, 2.
Cochran, W.D., Hatzes, A.P., Butler, R.P., Marcy, G.W.: 1997, *ApJ* **483**, 457.
Goldreich, P., Tremaine, S.: 1980, *ApJ* **241**, 425.
Gonzalez, G.: 1998, *A&A* **334**, 221.
Israelian, G., Santos, N.C., Mayor, M., Rebolo, R.: 2002, *A&A* , submitted.
Israelian, G., Santos, N.C., Mayor, M., Rebolo, R.: 2001, *Nat* **411**, 163.
Jorissen, A., Mayor, M., Udry, S.: 2001, *A&A* **379**, 992.
Laws, C., Gonzalez, G.: 2001, *ApJ* **553**, 405.
Lin, D.N.C., Ida, S.: 1997, *ApJ* **477**, 781.
Lin, D.N.C., Bodenheimer, P., Richardson, D.C.: 1996, *Nat* **380**, 606.
Mazeh, T., Shaham, J.: 1979, *A&A* **77**, 145.
Mayor, M., Queloz, D.: 1995, *Nat* **378**, 355.
Murray, N., Paskiwitz, M., Holman, M.: 2002, *ApJ* **565**, 608.
Murray, N., Hansen, B., Holman, M., Tremaine, S.: 1998, *Science* **279**, 69.
Naef, D., Latham, D.W., Mayor, M., Mazeh, T., Beuzit, J.L., Drukier, G.A., Perrier-Bellet, C., Queloz, D., Sivan, J.P., Torres, G., Udry, S., Zucker, S.: 2001, *A&A* **279**, 69.
Pinsonneault, M.H., DePoy, D.L., Coffee, M.: 2001, *ApJL* **556**, 59.
Pollack, J.B., Hubickyj, O., Bodenheimer, P., Lissauer, J.J., Podolak, M., Greenzweig, Y.: 1996, *Icarus* **124**, 62.
Rasio, F.A., Ford, E.B.: 1996, *Science* **274**, 954.
Reddy, B.E., Lambert, D.L., Laws, C., Gonzalez, G., Covey, K.: 2002, *MNRAS* **335**, 1005.
Santos, N.C., Garcia Lopez, R.J., Israelian, G., Mayor, M., Rebolo, R., Garcia-Gil, A., Perez de Taoro, M.R., Randich, S.: 2002, *A&A* **386**, 1028.
Santos, N.C., Israelian, G., Mayor, M., Rebolo, R., Udry, S.: 2002, *A&A* **398**, 363.
Santos, N.C., Mayor, M., Naef, D., Pepe, F., Queloz, D., Udry, S., Burnet, M., Clausen, J.V., Helt, B.E., Olsen, E.H., Pritchard, J.D.: 2002, *A&A* , **392**, 215.

Continuum Excess Emission in Young Low Mass Stars

Daniel F. M. Folha

Centro de Astrofísica da Universidade do Porto, Rua das Estrelas, 4150-762 Porto, Portugal

2002 November 4

Abstract. The amount of excess emission in the near-infrared spectrum of classical T Tauri stars (CTTS) is higher than predicted by current models. This lack of understanding about the CTTS systems impinges directly on our ability to achieve a deeper knowledge of the final stages of low mass stars formation, of the pre-main sequence history of the Sun and even on the processes that lead to planet formation.

Here I present a portion of the "Unsolved Universe" of the CTTS systems and some "Challenges for the Future" in this field.

Keywords: stars: circumstellar matter, stars: formation, stars: pre-main sequence, infrared: general, infrared:stars

1. Introduction

T Tauri stars (TTS) are young pre-main sequence stars which are analogs of the pre-main sequence Sun. They are divided into two classes: Classical TTS (CTTS) and Weak TTS (WTTS). The current paradigm is that CTTS are actively accreting from a circumstellar disk which is disrupted in its inner region by a stellar magnetic field. From the disk inner radius matter follows the magnetic field lines down to the stellar photosphere at nearly free-fall velocities, where it forms a shock as it accretes onto the star (e.g. Camenzind, 1990; Königl, 1991; Shu et al., 1994). WTTS lack the accretion signatures and their activity is most likely explained as the result of a magnetic dynamo (e.g. Bouvier, 1990), i.e. it is solar like activity but in a much larger scale.

Within this scenario, one can find various sources of possible continuum emission in CTTS: the stellar photosphere, the accretion shock(s), the accretion disk, star spots, and a residual circumstellar envelope. In the context of TTS, when referring to excess emission, one means the sum of all non-photospheric continuum contributions.

The excess continuum emission observed in these stars at ultraviolet and optical wavelengths smaller than ~ 5500 Å has been successfully explained as resulting from the accretion shock (Calvet & Gullbring, 1998). Infrared broadband colours and millimetre emission of CTTS has been explained by emission from a flaring accretion disk (Kenyon & Hartmann, 1987; Meyer et al., 1997), with occasional inclusion of an additional circumstellar envelope component (Hartmann, 1995). From these studies, involving different wavelength regions, the origin of the excess emission in CTTS seemed well understood.

2. Near-Infrared Veiling

Veiling is a relative measure of the continuum excess emission. At a given wavelength, the veiling r_λ is defined as the ratio of the excess flux to the stellar photospheric flux. It can be measured in CTTS by comparing the strength of their photospheric absorption lines with those of appropriate non-accreting stars (e.g. Basri & Batalha, 1990; Hartigan et al., 1989; Guenther & Hessman, 1993). A very important characteristic of veiling measurements is that they provide an estimate of the amount of excess emission independent of reddening, hence avoiding the problems of poorly known extinction, which can hamper photometric attempts to derive the excess emission.

Folha & Emerson (1999) study the veiling at the near-infrared (NIR), using high resolution NIR spectroscopy, for a relatively large sample of CTTS (45 CTTS observed in the J-band and 31 CTTS observed in the K-band). Photospheric lines were identified in the spectra of 73% of the CTTS observed at J and in 71% at K. The distribution of the veiling measurements is shown in Figure 1. The average values for the veiling are $\langle r_J \rangle = 0.57$ and

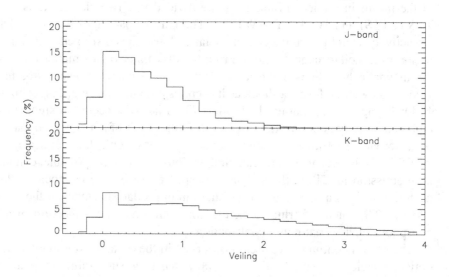

Figure 1. Distribution of the J-band (top panel) and K-band (bottom panel) veiling for the stars with detected photospheric lines.

$\langle r_k \rangle = 1.76$. For the remaining 27% of the stars observed at J and 29% at K only lower limits could be obtained for r_J and r_K. The average lower limits are $\langle r_J \rangle > 1.34$ and $\langle r_k \rangle > 2.4$.

While *J*-band veiling measurements are not found in the literature, *K*-band veiling have been measured for a sample of 14 CTTS by Johns-Krull & Valenti (2001). There are 11 CTTS common to the Folha & Emerson (1999) and Johns-Krull & Valenti (2001) samples, and the latter study corroborates the results found by the former.

3. Origin of NIR veiling

3.1. ACCRETION SHOCK

The accretion shock models (Calvet & Gullbring, 1998) that successfully explain the observed excess emission at UV and optical wavelengths smaller than about 5500 Å predict what the NIR veiling should be. Quoting Calvet & Gullbring (1998), "At NIR wavelengths the predicted veiling from the accretion column for typical CTTS parameters is nearly constant and ≤ 0.1. Only for the "continuum stars" are significant amounts of NIR veiling from the accretion column expected".

From Figure 1 it clear that the majority of CTTS in the Folha & Emerson (1999) have NIR veilings larger than 0.1. And only a very small number of those are "continuum stars" at optical wavelengths. Clearly, the accretion shock models do not provide an explanation for the observed NIR veiling in CTTS.

3.2. ACCRETION DISK

Meyer et al. (1997) model accretion disks around a typical CTTS and compute the expected emission spectrum. With this in hand and after averaging over an appropriate range of accretion rates, stellar masses and inclinations, they predict the veiling that should be observed at the *J*, *H*, *K* and *L* NIR bands. Their results are shown in the form of cumulative histograms for the veiling distributions.

When one compares the predicted cumulative veiling distributions with those that result from the distributions shown in Figure 1 one concludes that accretion disks with small ($< 2R_*$) inner hole sizes are typically needed to explain the observed veilings. However, the distributions shown in Figure 1 do not take into account the stars for which only lower limits were obtained for the veiling. Those distributions would certainly be enhanced for higher values of the veiling by including the latter stars. In fact, given such an enhancement an according to the Meyer et al. (1997) results, even disks without inner holes can be incapable of explaining the high veiling observed in the NIR spectra of many CTTS.

Johns-Krull & Valenti (2001) test observed *K*-band veilings in CTTS in the context of accretion disk models, by adopting the formalism of Chiang &

Goldreich (1997) and Chiang & Goldreich (1999) and implementing various modifications to these models. Namely, they consider smaller disks with internal heating, include radiation from the accretion shock, use improved grain emissivity, and experiment with different dust destruction temperatures, inner hole sizes and dust grain sizes. Johns-Krull & Valenti (2001) conclude that in general the predicted emission from current accretion disk models, constrained by the most recent estimates of mass accretion rates, is insufficient to reproduce the high K-band veiling measurements.

3.3. ALTERNATIVE EXPLANATIONS

That current accretion shock (one component models) and accretion disk models do not easily explain the moderate to high values of J- and K-band veiling is now well established. But is the NIR veiling closely related to the accretion process? In Figure 2 is plotted the J-band veiling versus the mass accretion rate derived from optical data (Hartigan et al., 1995; Gullbring et al., 1998). From this figure one sees that the J-band veiling correlates well with

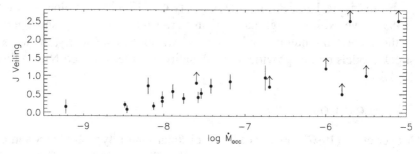

Figure 2. J-band veiling vs. mass accretion rate. Lower limits on the veiling are indicated by an arrow. Mass accretion rates for stars with veiling lower limits only are from Hartigan et al. (1995). Mass accretion rates for the remaining stars are from Gullbring et al. (1998).

the mass accretion rate. This correlation is further enhanced by noting that stars for which only lower limits on the veiling were computed tend to have high mass accretion rates. A similar correlation might be present between the K-band veiling and the mass accretion rate (plot not shown here), however the large uncertainties in the K veiling make a clear judgement difficult. The relationship between NIR veiling and mass accretion rate seems to point out that whatever the origin of the observed NIR veiling, and in particular of the J-band veiling, it is somehow influenced by accretion.

At this moment it is still unclear where the observed NIR excess emission discussed here originates. One possibility deserving further attention is that the excess arises in accretion shocks which are described by a superposition of accretion columns carrying different energy fluxes, instead of the single component model that explains the UV and blue wavelength observations.

In fact, Calvet & Gullbring (1998) note that veiling measurements in the red part of the spectrum for the CTTS BP Tau already show higher than expected excess emission.

Folha & Emerson (1999) speculate that continuum emission resulting from within the accretion flow itself may account for the high NIR veiling observed. Johns-Krull & Valenti (2001) elaborate further on this speculation and suggest that dust out of the plane of the accretion disk, and possibly in the magnetospheric flow itself, may be the source of the hitherto missing emission component.

4. Future

How well does one know the missing emission component? In fact, not very well at all. The NIR studies produced thus far that reveal its presence encompass very limited wavelength ranges. In addition, the data used was not flux calibrated nor simultaneous NIR photometry exists, hence not allowing one to find the flux calibrated spectra of the extra emission component. Furthermore, it is not yet clear whether this excess component can be traced into optical wavelengths. Gameiro et al. (2002) find what seems to be a steady increase in the veiling of the CTTS WY Ari (LkHα 264) from the red to the infrared (their Figure 5). This increase may be the optical signature of the excess emission component observed as J- and K-band veiling.

A full characterisation of the emission component giving rise to the observed high NIR veiling is crucial if one is to fully understand its origin. In order to achieve that goal it is necessary to derive flux calibrated spectra of the excess emission in CTTS with the longest wavelength coverage possible. In particular, one wants to determine the spectrum of the excess emission from the blue, where it has been convincingly shown that the accretion shock dominates, to the K-band, where emission from the accretion disk starts dominating. With that aim in sight, we (N. Calvet, V. Costa, D. Folha & J. Gameiro) have used SOFI on the ESO NTT and the Boller & Chievens spectrograph on the ESO 1.5m to obtain (quasi-)simultaneous optical (4400–7550 Å) and NIR (9800–11080 Å, J-, H- and K-band) flux calibrated spectra for a number of southern CTTS. We are currently working on the extraction of the excess emission spectrum of the observed stars. The unprecedented wavelength coverage of these data (for CTTS) makes it a valuable instrument for the study of the continuum excess emission in CTTS.

5. Concluding Remarks

A complete understanding of the CTTS systems is of major importance. The young Sun went through the T Tauri phase. Only looking into the CTTS systems can one be told the history of the pre-main sequence Sun and hence understand in detail how the Sun came to be as we see it today. The beginning of the process of planet formation most likely occurs during the T Tauri phase, inside the circumstellar (accretion) disks. Trilling et al. (2002) argue that at least 10% and perhaps as many as 80% of the solar-type stars possess giant planets during their pre-main sequence phase, i.e. during their T Tauri phase. If one wants to probe into the details of planet formation one will have to look at TTS. And then, a detailed knowledge of the various emission processes within those systems is certainly of outstanding importance.

Acknowledgements

The author would like to thank the organisation of JENAM 2002 for the invitation to present a highlight talk to a plenary session and the opportunity to produce this written contribution, as well as V. Costa for critically reading the manuscript. This work was financially supported by FCT through the "Subprograma Ciência e Tecnologia do 3º Quadro Comunitário de Apoio" and also by project POCTI/1999/FIS/34549, approved by FCT and POCTI, with funds from the European Union programme FEDER.

References

Basri, G., Batalha, C.: 1990, *ApJ* **363**, 654.
Bouvier, J.:1990, *AJ* **99**, 946.
Calvet, N. & Gullbring, E.: 1998, *ApJ* **509**, 802.
Camenzind, M.: 1990, *Reviews in Modern Astronomy* **3**, 234.
Chiang, E.I., Goldreich, P.: 1997, *ApJ* **490**, 368.
Chiang, E.I., Goldreich, P.: 1999, *ApJ* **519**, 279.
Folha, D.F.M., Emerson, J.P.: 1999, *A&A* **352**, 517.
Gameiro, J.F., Folha, D.F.M., Costa, V.M.: 2002, *A&A* **388**, 504.
Guenther, E., Hessman, F.V.: 1993, *A&A* **268**, 192.
Gullbring, E., Hartmann, L., Briceño, C., Calvet, N.: 1998, *ApJ* **492**, 323.
Hartigan, P., Hartmann, L., Kenyon, S., Hewett, R., Stauffer, J.: 1989, *ApJS* **70**, 899.
Hartigan, P., Edwards, S., Ghandour, L.: 1995, *ApJ* **452**, 736.
Hartmann, L.W.: 1995, *Ap&SS* **224**, 3.
Johns-Krull, C.M., Valenti, J.A.: 2001, *ApJ* **561**, 1060.
Kenyon, S.J., Hartmann, L.W.: 1987, *ApJ* **323**, 714.
Königl, A.: 1991, *ApJ* **370**, L39.
Meyer, M.R., Calvet, N., Hillenbrand, L.A.: 1997, *AJ* **114**, 288.
Shu, F., Najita, J., Ostriker, E., Wilkin, F., Ruden, S., Lizano, S.: 1994, *ApJ* **429**, 781.
Trilling, D.E., Lunine, J.I., Benz, W.: 2002, *A&A* **394**, 241.

The Formation and Evolution of Field Massive Galaxies

Andrea Cimatti
INAF - Osservatorio Astrofisico di Arcetri, Italy

2002 November 5

Abstract. The problem of the formation and evolution of field massive galaxies is briefly reviewed from an observational perspective. The motivations and the characteristics of the K20 survey are outlined. The redshift distribution of $K_s < 20$ galaxies, the evolution of the rest-frame K_s-band luminosity function and luminosity density to $z \sim 1.5$, the nature and the role of the red galaxy population are presented. Such results are compared with the predictions of models of galaxy evolution.

Keywords: galaxies; cosmology

1. Introduction

Despite the recent developments in observational cosmology, one of the main unsolved issues remains how and when the present-day massive elliptical galaxies ($\mathcal{M}_{stars} > 10^{11}$ $M_\odot$) built up and what type of evolution characterized their growth across the cosmic time.

There are two main proposed scenarios. In the first, such systems formed at high redshifts (e.g. $z > 2-3$) through a "monolithic" collapse accompanied by a violent burst of star formation, then followed by a passive and pure luminosity evolution (PLE) of the stellar population to nowadays (Eggen, Lynden-Bell & Sandage, 1962; Tinsley, 1972; Larson, 1974; van Albada, 1982). Such a scenario makes some critical and rigid predictions that can be tested with the observations: *(i)* the comoving number density of massive spheroids should be conserved through cosmic times, *(ii)* massive galaxies should evolve only in luminosity, *(iii)* old passively evolving spheroids should exist at least up to $z \sim 1-1.5$, *(iv)* there should be a population of progenitors at $z > 2-3$ characterized by large amounts of gas (and dust) and strong star formation rates in order to be compatible with the rapid formation scenario and with the properties (e.g. masses, ages, metallicities) of the present-day "fossils" resulting from that formation process.

In a diametrically opposed scenario, massive spheroids formed at later times through a slower process of hierarchical merging of smaller galaxies (e.g. White & Rees, 1978; Kauffman, White, & Guiderdoni, 1993; Kauffmann, 1996) characterized by moderate star formation rates, thus reaching the final masses in more recent epochs (e.g. $z < 1-1.5$) (e.g. Baugh et al., 1996, 1998; Cole et al., 2000; Baugh et al., 2002). As a consequence, the hierarchical merging models (HMMs) predict that massive systems should

M.J.P.F.G. Monteiro (ed.), The Unsolved Universe: Challenges for the Future, 31-48.
© 2003 *Kluwer Academic Publishers.*

be very rare at $z \sim 1$, with the comoving density of $\mathcal{M}_{stars} > 10^{11}$ $M_\odot$ galaxies decreasing by almost an order of magnitude from $z \sim 0$ to $z \sim 1$ (Baugh et al., 2002; Benson et al., 2002).

Several observations were designed over the recent years in order to test such two competing models.

One possibility is to search for the starburst progenitors expected at $z >$ 2−3 in the "monolithic"+PLE scenario. In this respect, submm and mm continuum surveys unveiled a population of high-z dusty starbursts which may represent the ancestors of the present-day massive galaxies (see Blain et al., 2002, for a recent review).

The other possibility is to search for passively evolving spheroids to the highest possible redshifts and to study their properties both in clusters and in the field. This latter approach provided so far controversial results.

Because of their color evolution, fundamental plane and stellar population properties, cluster ellipticals are now generally believed to form a homogeneous population of old systems formed at high redshifts (e.g. Stanford et al., 1998; see also Renzini, 1999; Renzini & Cimatti, 1999; Peebles, 2002, for recent reviews).

However, the question of field spheroids is still actively debated. It is now established that old, passive and massive systems exist in the field out to $z \sim 1.5$ (e.g. Spinrad et al., 1997; Stiavelli et al., 1999; Waddington et al., 2002), but the open question is what are their number density and physical/evolutionary properties with respect to the model predictions.

Some surveys based on color or morphological selections found a deficit of $z > 1-1.4$ elliptical candidates (e.g. Kauffmann et al., 1996; Zepf, 1997; Franceschini et al., 1998; Barger et al., 1999; Rodighiero et al., 2001; Smith et al., 2002; Roche et al., 2002), whereas others did not confirm such result out to $z \sim 1-2$ (e.g. Totani & Yoshii, 1998; Benitez et al., 1999; Daddi et al., 2000b; Im et al., 2002; Cimatti et al., 2002a). Part of the discrepancies can be ascribed to the strong clustering (hence field-to-field variations) of the galaxies with the red colors expected for high-z elliptical candidates (Daddi et al., 2000a).

Other approaches made the picture even more controversial. For instance, Menanteau et al. (2001) found that a fraction of morphologically selected field spheroidals show internal color variations incompatible with a traditional PLE scenario and stronger than cluster spheroidals at the same redshifts. Similar results have been obtained with photometric, spectroscopic and fundamental plane studies of field ellipticals to $z \sim 0.7-1$ (e.g. Kodama et al., 1999; Schade et al., 1999; Treu et al., 2002). Such observations suggest that, despite the mass of massive spheroids seems not to change significantly from $z \sim 1$ to $z \sim 0$ (Brinchmann & Ellis, 2000), field early-type systems at $z \sim 0.5-1$ do not form an entirely homogeneous population, some looking consistent with the PLE scenario, whereas others with signatures of recent secondary episodes of star formation (see also Ellis, 2000, for a review).

A more solid and unbiased approach is to investigate the evolution of massive galaxies by means of spectroscopic surveys of field galaxies selected in the K-band (e.g. Broadhurst et al., 1992), and to push the study of massive systems to $z > 1$. Since the rest-frame optical and near-IR light is a good tracer of the galaxy *stellar* mass (Gavazzi et al., 1996), K-band surveys provide the important possibility to select galaxies according to their mass up to $z \sim 2$. The advantages of the K-band selection also include the small k-corrections with respect to optical surveys (which are sensitive to the star formation activity rather than to the stellar mass), and the minor effects of dust extinction. Once a sample of faint field galaxies has been selected in the K-band, deep spectroscopy with 8-10m class telescopes can then be performed to shed light on their nature and on their redshift distribution. Several spectroscopic surveys of this kind have been and are being performed (e.g. Cowie et al., 1996; Cohen et al., 1999; Stern et al., 2001; see also Drory et al., 2001, although mostly based on photometric redshifts).

In this paper, the main results obtained so far with a new spectroscopic survey for K-selected field galaxies are reviewed, concentrating on the redshift distribution, the evolution of the near-IR luminosity function and luminosity density, the very red galaxy population, and on the comparison with the predictions of the most recent scenarios of galaxy formation and evolution. $H_0=70$ km s^{-1} Mpc^{-1}, $\Omega_m=0.3$ and $\Omega_\Lambda=0.7$ are adopted.

2. The K20 survey

Motivated by the above open questions, we started an ESO VLT Large Program (dubbed "K20 survey") based on 17 nights distributed over two years (1999-2000) (see Cimatti et al., 2002c, for details).

The prime aim of such a survey was to derive the redshift distribution and spectral properties of 546 K_s-selected objects with the *only* selection criterion of $K_s < 20$ (Vega). Such a threshold is critical because it selects galaxies over a broad range of masses, i.e. $\mathcal{M}_{stars}>10^{10}$ M$_\odot$ and $\mathcal{M}_{stars}>4\times10^{10}$ M$_\odot$ for $z=0.5$ and $z=1$ respectively (according to the mean $\mathcal{M}_{stars}/L$ ratio in the local universe and adopting Bruzual & Charlot 2000 spectral synthesis models with a Salpeter IMF). The $K_s < 20$ selection has also the observational advantage that most galaxies have magnitudes still within the limits of optical spectroscopy of 8m-class telescopes ($R < 25$).

The targets were selected from K_s-band images (ESO NTT+SOFI) of *two independent fields* covering a total area of 52 arcmin^2. One of the fields is a sub-area of the Chandra Deep Field South (CDFS; Giacconi et al., 2001). Optical multi-object spectroscopy was made with the ESO VLT UT1 and

UT2 equipped with FORS1 and FORS2. A fraction of the sample was also observed with near-IR spectroscopy with VLT UT1+ISAAC in order to attempt to derive the redshifts of the galaxies which were too faint for optical spectroscopy and/or expected to be in a redshift range for which no strong features fall in the observed optical spectral region (e.g. $1.5 < z < 2.0$). In addition to spectroscopy, $UBVRIzJK_s$ imaging was also available for both fields, thus providing the possibility to estimate photometric redshifts for all the objects in the K20 sample, to optimize them through a comparison with the spectroscopic redshifts and to assign a reliable photometric redshift to the objects for which it was not possible to derive the spectroscopic z. The overall spectroscopic redshift completeness is 94%, 92%, 87% for $K_s < 19.0$, 19.5, 20.0 respectively. The overall redshift completeness (spectroscopic + photometric redshifts) is 98%.

The K20 survey represents a significant improvement with respect to previous surveys for faint K-selected galaxies (e.g. Cowie et al., 1996; Cohen et al., 1999) thanks to its larger sample, the coverage of two independent fields (thus reducing the cosmic variance effects), the availability of optimized photometric redshifts, and the spectroscopic redshift completeness, in particular for the reddest galaxies.

3. The redshift distribution of $K_s < 20$ galaxies

The observed differential and cumulative redshift distributions for the K20 sample are presented in Figure 1 (see Cimatti et al., 2002b), together with the predictions of different scenarios of galaxy formation and evolution, including both hierarchical merging models (HMMs) from Menci et al. (2002, M02), Cole et al. (2000, C00), Somerville et al. (2001, S01), and pure luminosity evolution models (PLE) based on Pozzetti et al. (1996, 1998 PPLE) and Totani et al. (2001, TPLE). The redshift distribution can be retrieved from http://www.arcetri.astro.it/~k20/releases. The spike at $z \sim 0.7$ is due to two clusters (or rich groups) at $z=0.67$ and $z=0.73$. The median redshift of $N(z)$ is $z_{med}=0.737$ and $z_{med}=0.805$, respectively with and without the two clusters being included. Without the clusters, the fractions of galaxies at $z > 1$ and $z > 1.5$ are 138/424 (32.5%) and 39/424 (9.2%) respectively. The high-z tail extends beyond $z=2$. The contribution of objects with only a photometric redshift becomes relevant only for $z > 1.5$. The fractional cumulative distributions displayed in Figure 1 (bottom panels) were obtained by removing the two clusters mentioned above in order to perform a meaningful comparison with the galaxy formation models which do not include clusters (PLE models), or are averaged over very large volumes, hence diluting the effects of redshift spikes (HMMs). No best tuning of the models was attempted in this comparison, thus allowing an unbiased *blind test* with the K20 observational data. The model predicted $N(z)$ are normalized to the K20 survey sky area.

Figure 1a shows a fairly good agreement between the observed $N(z)$ distribution and the PLE models (with the exception of PPLE with Salpeter IMF), although such models slightly overpredict the number of galaxies at $z \gtrsim 1.2$. However, if the photometric selection effects present in the K20 survey (Cimatti et al. 2002b) are taken into account, the PLE models become much closer to the observed $N(z)$ thanks to the decrease of the predicted high-z tail. According to the Kolmogorov-Smirnov test, the PLE models are acceptable at 95% confidence level, with the exception of the PPLE model with Salpeter IMF.

On the other side, all the HMMs underpredict the median redshift (i.e. z_{med}=0.59, 0.70 and 0.67 for the C00, M02 and S01 models respectively), overpredict the total number of galaxies with $K_s < 20$ by factors up to $\sim$50% as well as the number of galaxies at $z < 0.5$, and underpredict the fractions of $z > 1-1.5$ galaxies by factors of $2-4$ (Figure 1b). Figure 1b (bottom panels) illustrates that in the fractional cumulative distributions the discrepancy with observations appears systematic at all redshifts. The Kolmogorov-Smirnov test shows that all the HMMs are discrepant with the observations at $> 99\%$ level. The inclusion of the photometric biases exacerbates this discrepancy, as shown in Figure 1b (right panels) for the M02 model (the discrepancy for the C00 and S01 models becomes even stronger). The deficit of high-redshift objects is well illustrated by Figure 2, where the PPLE model is capable to reproduce the cumulative *number* distribution of galaxies at $1 < z < 3$ within 1-2σ, whereas the M02 model is always discrepant at $\geq 3\sigma$ level (up to $>$ 5σ for $1.5 < z < 2.5$). This conclusion is not heavily based on the objects with only photometric redshifts estimates, as the mere presence of 7 galaxies with spectroscopic redshift $z > 1.6$ is already in substantial contrast with the predictions by HMMs of basically no galaxies with $K_s < 20$ and $z > 1.6$.

4. The evolution of the luminosity function

The luminosity function (LF) of galaxies has been estimated in the rest-frame K_s-band and in three redshift bins which avoid the clusters at $z \sim 0.7$ (z_{mean}=0.5,1,1.5; see Figure 3) (Pozzetti et al. 2003), using both the $1/V_{max}$ (Schmidt 1968; Felten 1976) and the STY (Sandage, Tammann & Yahil 1979) formalisms. The LF observed in the first two redshift bins is fairly well fit by Schechter functions. A comparison with the local K_s-band LF of Cole et al. (2001) shows a mild *luminosity* evolution of LF(z) out to $z = 1$, with a brightening of about -0.5 magnitudes from $z = 0$ to $z = 1$ (Figure 3). Similar results have been found by Drory et al. (2001), Cohen (2002), Bolzonella et al. (2002) and Miyazaki et al. (2002) (see also Cowie et al., 1996).

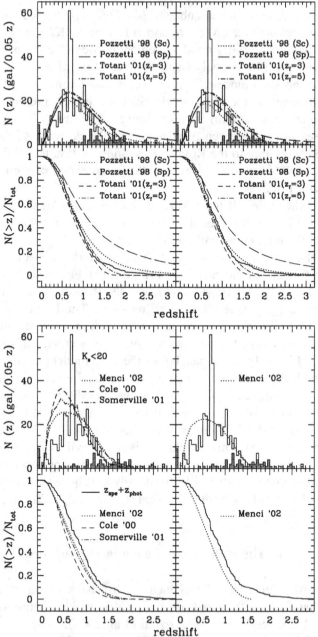

Figure 1. (**a**) – *Top panels:* the observed differential $N(z)$ for $K_s < 20$ (histogram) compared with the PLE model predictions. *Bottom panels:* the observed fractional cumulative redshift distribution (continuous line) compared with the same models. The shaded histogram shows the contribution of photometric redshifts. The bin at $z < 0$ indicates the 9 objects without redshift. The *left* and *right* panels show the models without and with the inclusion of the photometric selection effects respectively. Sc and Sp indicate Scalo and Salpeter IMFs respectively. (**b**) – same as Figure 1a, but compared with the HMM predictions. *Right* panels: the M02 model with the inclusion of the photometric selection effects.

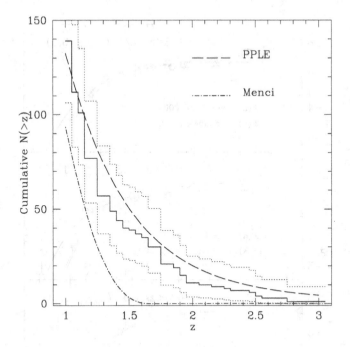

Figure 2. The observed cumulative *number* of galaxies between $1 < z < 3$ (continuous line) and the corresponding poissonian $\pm 3\sigma$ confidence region (dotted lines). The PPLE (Scalo IMF) and the M02 models are corrected for the photometric biases.

The study of the LF by galaxy spectral or color types shows that red early-type galaxies dominate the bright-end of the LF already at $z \sim 1$, and that their number density shows only a small decrease from $z \sim 0$ to $z \sim 1$ (Pozzetti et al., 2003). This is consistent with the independent study of Im et al. (2002) based on morphologically selected spheroidals.

Figure 4 shows a comparison of the observed luminosity function with PLE and HMM predictions. The PLE models describe reasonably well the shape and the evolution of the luminosity function up to the highest redshift bin, $z_{mean} = 1.5$, with no evidence for a strong decline of the most luminous systems (with $L > L^*$). This is in contrast, especially in the highest redshift bin, with the prediction by the HMMs of a decline in the number density of luminous (i.e. massive) systems with redshift. Moreover, hierarchical merging models (namely M02 and C00) result in a significant overprediction of faint, sub–L^* galaxies at $0 < z < 1.3$. This problem, also hinted by the comparison of $N(z)$ between models and data, is probably related to the so called "satellite problem" (e.g. Primack, 2002).

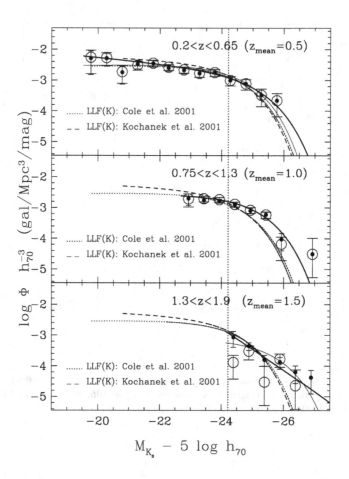

Figure 3. The rest-frame K_s-band Luminosity Function in three redshift bins. Data points derive from $1/V_{max}$ analysis. Solid curves: the Schechter fits derived from maximum likelihood analysis (thin solid lines are the fit assuming local α parameter). Dotted and dashed curves: the local K_s-band LFs of Cole et al. (2001) and Kochanek et al. (2001) respectively. The vertical dotted line indicates the local M^* of Cole et al. (2000). Open circles: spectroscopic redshifts, filled circles: spectroscopic + photometric redshifts.

However, it is interesting to note that at $z \sim 1$ the HMMs seem not to be in strong disagreement with the observations relative to the bright end of the galaxy luminosity function (with the possible exception of the highest luminosity point). Thus, the key issue is to verify whether the bright $L > L^*$ galaxies in the K20 survey have the same nature of the luminous galaxies predicted by the HMMs, in particular for their mass to light ratios ($\mathcal{M}_{stars}/L$).

Figure 5 compares the $R - K_s$ colors and luminosity distributions of galaxies with $0.75 < z < 1.3$ (a bin dominated by spectroscopic redshifts) as ob-

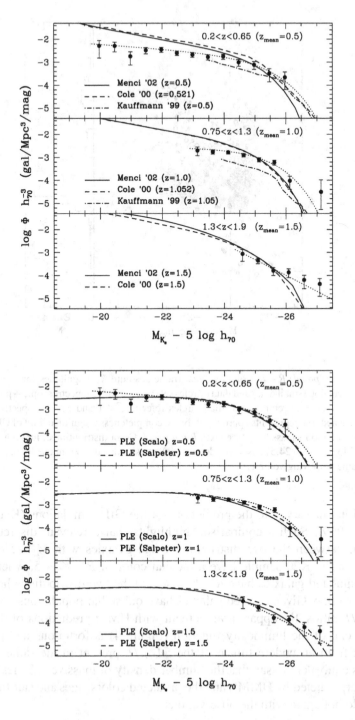

Figure 4. Left: the K_s-band LF compared to hierarchical merging model predictions. Right: the K_s-band LF compared to PLE (PPLE) model predictions. Dotted curves are the Schechter best fits to the observed LFs (spectroscopic + photometric redshifts).

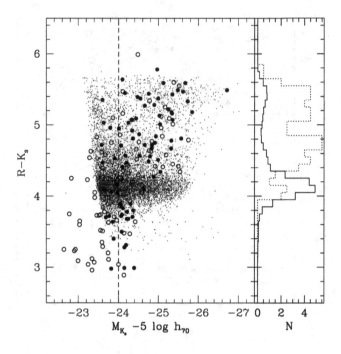

Figure 5. Left panel: $R - K_s$ colors vs. rest-frame absolute K_s magnitudes for $z = 1.05$ GIF simulated catalog (small dots) and data (circles) at $0.75 < z < 1.3$ (spectroscopic + photometric redshifts; $z_{mean} = 1$) (empty and filled circles refer to $z < 1$ and $z > 1$ respectively). The vertical dashed line represents approximately the completeness magnitude limit of GIF catalog corresponding to its mass limit (see text). *Right panel:* Color distribution of luminous galaxies ($M_{K_s} - 5 \log h_{70} < -24.5$) observed (dotted line) and simulated (continuous line), normalized to the same comoving volume.

served in our survey to the predictions of the GIF[1] simulations (Kauffmann et al., 1999). Such a comparison highlights that a relevant discrepancy is present between the two distributions: real galaxies with $M_K - 5\log h_{70} < -24.5$ in the K20 sample have a median color of $R - K_s \sim 5$, whereas the GIF simulated galaxies have $R - K_s \sim 4$, and the two distributions have very small overlap. Given that red galaxies have old stellar populations and higher $\mathcal{M}_{stars}/L$ ratios, the apparent agreement with HMM predictions of the $z \sim 1$ bright end of the luminosity function (Figure 4) is fortuitous and probably results from an underestimate of the $\mathcal{M}_{stars}/L$ present in the same models. This is equivalent to say that the number density of massive galaxies at $z \sim 1$ is underpredicted by HMMs, and the predicted colors, ages and star formation rates do not agree with the observations.

---

[1] http://www.mpa-garching.mpg.de/GIF/

5. The evolution of the luminosity density

Tracing the integrated cosmic emission history of the galaxies at different wavelengths offers the prospect of an empirical determination of the global evolution of the galaxy population. Indeed it is independent of the details of galaxy evolution and depends mainly on the star formation history of the universe (Lilly et al., 1996; Madau, Pozzetti & Dickinson, 1998). Attempts to reconstruct the cosmic evolution of the comoving luminosity density have been made previously mainly in the UV and optical bands, i.e. focusing on the star formation history activity of galaxies (Lilly et al., 1996; Cowie et al., 1999).

Our survey offers for the first time the possibility to investigate it in the near-IR using a LF extended over a wide range in luminosity, thus providing new clues on the global evolution of the stellar mass density (Pozzetti et al., 2003). Using the local luminosity density at $z \sim 0$ as derived from Cole et al. (2001) complemented with the estimates at higher redshifts based on the K20 survey, it is found that the rest-frame K_s-band luminosity density up to $z \sim 1.3$ is well represented by a power law with $\rho_\lambda(z) = \rho_\lambda(z=0)(1+z)^\beta$, with $\beta = 0.37$. Compared to the optical (rest-frame UV-blue) bands, the near-IR luminosity density evolution is much slower ($\beta = 3.9 - 2.7$ from 0.28 to 0.44 μm by Lilly et al. (1996), and $\beta = 1.5$ at 0.15-0.28 μm by Cowie et al. (1999), for $\Omega_m = 1$). The slow evolution of the observed K_s-band luminosity density suggests that the stellar mass density should also evolve slowly at least up to $z \sim 1.3$. This is in agreement with a recent analysis by Bolzonella et al. (2002) (see also Cowie et al., 1996; and Brinchmann & Ellis, 2000). The analysis of the stellar mass function and its cosmic evolution is in progress and will be presented elsewhere.

6. Extremely Red Objects (EROs)

Extremely Red Objects (EROs, $R - K > 5$) are critical in the context of galaxy formation and evolution because their colors allow to select old and passively evolving galaxies at $z > 0.9$.

For a fraction of EROs (70% to $K_s < 19.2$) present in the K20 sample it was possible to derive a spectroscopic redshift and a spectral classification (Cimatti et al., 2002a). Two classes of galaxies at $z \sim 1$ contribute nearly equally to the ERO population: old stellar systems with no signs of star formation, and dusty star-forming galaxies.

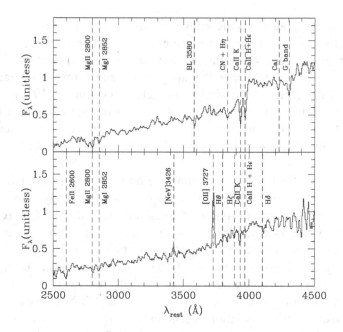

Figure 6. The average rest-frame spectra (smoothed with a 3 pixel boxcar) of old passively evolving (top; $z_{mean} = 1.000$) and dusty star-forming EROs (bottom; $z_{mean} = 1.096$) with $K_s \leq 20$ (Cimatti et al., 2002a).

6.1. OLD EROS

The colors and spectral properties of old EROs are consistent with ≥ 3 Gyr old passively evolving stellar populations (assuming solar metallicity and Salpeter IMF), requiring a formation redshift $z_f > 2.4$. The number density is $6.3 \pm 1.8 \times 10^{-4} h^3 \text{Mpc}^{-3}$ for $K_s < 19.2$, consistent with the expectations of PLE models for passively evolving early-type galaxies with similar formation redshifts (Cimatti et al., 2002a). HMMs predict a significant deficit of such old red galaxies at $z \sim 1$, ranging from a factor of ~ 3 (Kauffmann et al., 1999) to a factor of ~ 5 (Cole et al., 2000). Preliminary analysis of recent HST+ACS imaging shows that old EROs have indeed spheroidal morphologies with surface brightness profiles typical of elliptical galaxies.

6.2. DUSTY STAR-FORMING EROS

The spectra of star-forming EROs suggest a dust reddening of $E(B-V) \sim$ 0.5–1 (adopting the Calzetti, 2001, extinction law), implying typical star-formation rates of 50-150 $M_\odot \text{yr}^{-1}$, and a significant contribution ($> 20 -$

30%) to the cosmic star-formation density at $z \sim 1$ (see also Smail et al., 2002). A recent analysis based on their X-ray emission provided a similar estimate of the SFRs (Brusa et al., 2002).

The comoving density of dusty EROs is again $\sim 6 \times 10^{-4}$ $h^3 Mpc^{-3}$ at $K_s < 19.2$. The GIF simulations (Kauffmann et al., 1999) predict a comoving density of red galaxies with $SFR > 50$ $M_\odot yr^{-1}$ that is a factor of 30 lower than the observed density of dusty EROs.

Such moderate SFRs suggest that the far-infrared luminosities of dusty star-forming EROs are generally below $L_{FIR} \sim 10^{12}$ $L_\odot$, and would then explain the origin of the low detection rate of EROs with $Ks < 20 - 20.5$ in submm continuum observations (e.g. Mohan et al., 2002; see also Smail et al., 2002). However, the fraction of dusty ultraluminous infrared systems may be higher in ERO samples selected at fainter Ks-band magnitudes (e.g. Wehner et al., 2002).

6.3. CLUSTERING

Taking advantage of the spectroscopic redshift information for the two ERO classes, we compared the relative 3D clustering in real space (Daddi et al., 2002). The comoving correlation lengths of dusty and old EROs are constrained to be $r_0 < 2.5$ and $5.5 < r_0 < 16$ h^{-1} Mpc comoving respectively, implying that old EROs are the main source of the ERO strong angular clustering. It is important to notice that the strong clustering measured for the old EROs is in agreement with the predictions of hierarchical clustering scenarios (Kauffmann et al., 1999).

7. Summary and discussion

The high level of completeness of the K20 survey and the relative set of results presented in previous sections provide new implications for a better understanding of the evolution of "mass-selected" field galaxies.

(1) The redshift distribution of $K_s < 20$ field galaxies has a median redshift of $z_{med} \sim 0.8$ and a high-z tail extended beyond $z \sim 2$. The current models of hierarchical merging do not match the observed median redshift because they significantly overpredict the number of low luminosity (hence low mass) galaxies at $z < 0.4 - 0.5$, and underpredict the fraction of objects at $z > 1 - 1.5$. Instead, the redshift distributions predicted by PLE models are in reasonable agreement with the observations. It is relevant to recall here that early predictions of the expected fraction of galaxies at $z > 1$ in a $K_s < 20$ sample indicated respectively $\approx 60\%$ and $\approx 10\%$ for a PLE case and for a (then) standard $\Omega_m = 1$ CDM model (Kauffmann & Charlot 1998). This version of PLE was then

ruled out by Fontana et al. (1999). The more recent PLE models and HMMs consistently show that for $z > 1$ the difference between the predictions of different scenarios is much less extreme. These results come partly from the now favoured ΛCDM cosmology which pushes most of the merging activity in hierarchical models at earlier times compared to τCDM and SCDM models with $\Omega_m = 1$ (structures form later in a matter-dominated universe, thus resulting in an even lower fraction of galaxies at high-z), and partly to different recipes for merging and star formation modes, which tend to narrow the gap between HMMs and the PLE case (e.g. Somerville et al. 2001; Menci et al., 2002). In this respect, the observed $N(z)$ provides an additional evidence that the universe is not matter-dominated ($\Omega_m < 1$).

(2) The rest-frame K_s-band luminosity function shows a mild luminosity evolution up to at least $z \sim 1$, with a brightening of about 0.5 magnitudes. Significant density evolution is ruled out up to $z \sim 1$. Current hierarchical merging models fail in reproducing the shape and evolutionary properties of the LF because they overpredict the number of sub-L^* galaxies and predict a substantial density evolution. PLE models are in good agreement with the observations up to $z \sim 1$.

(3) At odds with the HMMs, the bright-end of the LF at $z \sim 1$ is dominated by red and luminous (hence old and massive) galaxies.

(4) The rest-frame K_s-band luminosity density (hence the stellar mass density) evolves slowly up to $z \sim 1.3$.

(5) Old passive systems and dusty star-forming galaxies (both at $z \sim 1$) equally contribute to the ERO population with $K_s < 19.2$.

(6) The number, luminosities and ages of old EROs imply that massive spheroids formed at $z > 2.4$ and that were already fully assembled at $z \sim 1$, consistently with a PLE scenario.

(7) Dusty EROs allow to select (in a way complementary to other surveys for star-forming systems) a population of galaxies which contribute significantly to the cosmic star formation budget at $z \sim 1$.

(8) HMMs strongly underpredict the number of both ERO classes.

Overall, the results of the K20 survey show that galaxies selected in the K_s-band are characterized by little evolution up to $z \sim 1$, and that the observed properties can be successfully described by a PLE scenario. In contrast, HMMs fail in reproducing the observations because they predict a sort of "delayed" scenario where the assembly of massive galaxies occurs later than what is actually observed. We recall here that the discrepancies of HMMs in accounting for the properties of even $z = 0 \rightarrow \sim 1$ early-type galaxies have been already emphasized in the past (e.g., Renzini 1999; Renzini & Cimatti 1999). Moreover, among low-redshift galaxies there appears to be a clear anti-correlation of the specific star formation rate with galactic mass (Gavazzi et al., 1996; Boselli et al., 2001), the most massive galaxies being "old", the low-mass galaxies being instead dominated by young stellar populations. This is just the opposite than expected in the traditional HMMs, where the most massive galaxies are the last to form. The same anti-correlation is observed in the K20 survey at $z \sim 1$.

It is important to stress here that the above results do not necessarily mean that the whole framework of hierarchical merging of CDM halos is under discussion. For instance, the strong clustering of old EROs and the clustering evolution of the K20 galaxies (irrespective of colors) seem to be fully consistent with the predictions of CDM models of large scale structure evolution (Daddi et al., 2001; Firth et al., 2002; Daddi et al. in preparation).

It is also important to stress that the K20 survey allows to perform tests which are sensitive to the evolutionary "modes" of galaxies rather than to their formation mechanism. This means that merging, as the galaxy main formation mechanism, is not ruled out by the present observations. Also, it should be noted that PLE models are not a physical alternative to the HMMs, but rather tools useful to parameterize the evolution of galaxies under three main assumptions: high formation redshift, conservation of number density through cosmic times, passive luminosity evolution of the stellar populations.

Thus, if we still accept the ΛCDM scenario of hierarchical merging of dark matter halos as the *basic framework for structure and galaxy formation*, the observed discrepancies highlighted by the K20 survey may be ascribed to how the *baryon assembly* is treated and, in particular, to the heuristic algorithms adopted for the star formation processes and their feedback, both within individual galaxies and in their environment. Our results suggest that HMMs should have galaxy formation in a CDM dominated universe to closely mimic the old-fashioned *monolithic collapse* scenario. This requires enhancing merging and star formation in massive halos at high redshift (say, $z \gtrsim 2 - 3$), while in the meantime suppressing star formation in low-mass halos. For instance, Granato et al. (2001) suggested the strong UV radiation feedback from the AGN activity during the era of supermassive black hole formation to be responsible for the suppression of star formation in low-mass halos, hence imprinting a "anti-hierarchical" behavior in the baryonic component. The same effect may well result from the feedback by the starburst activity itself (see also Ferguson & Babul 1998).

In summary, the redshift distribution of $K_s < 20$ galaxies, together with the space density, nature, and clustering properties of the ERO population, and the redshift evolution of the rest-frame near-IR luminosity function and luminosity density provide a new set of observables on the galaxy population in the $z \sim 1 - 2$ universe, thus bridging the properties of $z \sim 0$ galaxies with those of Lyman-break and submm/mm-selected galaxies at $z \geq 2$–3. This set of observables poses a new challenge for theoretical models to properly reproduce.

Deeper spectroscopy coupled with HST+ACS imaging and SIRTF photometry will allow us to derive additional constraints on the nature and evolution of massive stellar systems out to higher redshifts.

Acknowledgements

The K20 survey team includes: S. Cristiani (INAF-Trieste), S. D'Odorico (ESO), A. Fontana (INAF-Roma), E. Giallongo (INAF-Roma), R. Gilmozzi (ESO), N. Menci (INAF-Roma), M. Mignoli (INAF-Bologna), F. Poli (University of Rome), A. Renzini (ESO), P. Saracco (INAF-Brera), J. Vernet (INAF-Arcetri), and G. Zamorani (INAF-Bologna).

We are grateful to C. Baugh, R. Somerville and T. Totani for providing their model predictions. AC warmly acknowledges Jim Peebles and Mark Dickinson for useful and stimulating discussions.

References

Barger, A.J., et al.: 1999, *AJ* **117**, 102

Baugh, C.M., Cole, S., Frenk, C.S.: 1996, *MNRAS* **283**, 1361

Baugh, C.M., Cole, S., Frenk, C.S., Lacey, C.G.: 1998, *ApJ* **498**, 504

Baugh, C.M., et al.: 2002, in *The Mass of Galaxies at Low and High Redshift*, Venice 2001, R. Bender, A. Renzini (eds), 91 [astro-ph/0203051]

Benitez, N., et al.: 1999, *ApJ* **515**, L65

Benson, A.J., Ellis, R.S., Menanteau, F.: 2002, *MNRAS* **336**, 564

Blain, A.W., Smail, I., Ivison, R.J., Kneib, J.-P., Frayer, D.T.: 2002, *Physics Reports* **369**, 111

Bolzonella, M., et al.: 2002, *A&A* **395**, 443

Boselli, A., Gavazzi, G., Donas, J., Scodeggio, M.: 2001, *AJ* **121**, 753

Brinchmann, J., Ellis, R.S.: 2000, *ApJ* **536**, L77

Broadhurst, T., Ellis, R.S., Grazebrook, K.: 1992, *Nat* **355**, 55

Brusa, M., et al.: 2002, *ApJL* **581**, 89

Calzetti, D.: 2001, *PASP* **113**, 1449

Cimatti, A., Daddi, E., Mignoli, M., et al.: 2002a, *A&A* **381**, L68

Cimatti, A., Pozzetti, L., Mignoli, M., et al.: 2002b, *A&A* **391**, L1

Cimatti, A., Mignoli, M., Daddi, E., et al.: 2002c, *A&A* **392**, 395

Cohen, J.G., Blandford, R., Hogg, D.W., et al.: 1999, *ApJ* **512**, 30

Cohen, J.G.: 2002, *ApJ* **567**, 672

Cole, S., Lacey, C.G., Baugh, C.M., Frenk, C.S.: 2000, *MNRAS* **319**, 168

Cole, S., Norberg, P., Baugh, C.M., et al.: 2001, *MNRAS* **326**, 255

Cowie, L.L., Songaila, A., Hu, E.M., Cohen, J.G.: 1996, *AJ* **112**, 839

Cowie, L.L., Songaila, A., Barger, A.J.: 1999, *AJ* **118**, 603

Daddi, E., et al.: 2000a, *A&A* **361**, 535

Daddi, E., Cimatti, A., Renzini, A.: 2000b, *A&A* **362**, L45

Daddi, E., Broadhurst, T., Zamorani, G., et al.: 2001, *A&A* **376**, 825

Daddi, E., Cimatti, A., Broadhurst, T., et al.: 2002, *A&A* **384**, L1 (Paper II)

Drory, N., Bender, R., Snigula, J., et al.: 2001, *ApJ* **562**, L111

Eggen, O.J., Lynden-Bell, D., Sandage, A.: 1962, *ApJ* **136**, 748

Ellis, R.S.: 2000, in *XIth Canary Islands Winter School of Astrophysics "Galaxies at High Redshift"*, [astro-ph/0102056]

Felten, J.E.: 1976, *ApJ* **207**, 700

Ferguson, H.C., Babul, A.: 1998, *MNRAS* **296**, 585

Firth, A.E., Somerville, R.S., McMahon, R.G., et al.: 2002, *MNRAS* **332**, 617

Fontana, A., Menci, N., D'Odorico, S., et al.: 1999, *MNRAS* **310**, L27

Franceschini, A., et al.: 1998, *ApJ* **506**, 600

Gavazzi, G., Pierini, D., Boselli, A.: 1996, *A&A* **312**, 397

Giacconi, R., Rosati, P., Tozzi, P., et al.: 2001, *ApJ* **551**, 624

Granato, G.L., Silva, L., Monaco, P., et al.: 2001, *MNRAS* **324**, 757

Im, M., et al.: 2002, *ApJ* **571**, 136

Kauffmann, G., White, S.D.M., Guiderdoni, B.: 1993, *MNRAS* **264**, 201

Kauffmann, G.: 1996, *MNRAS* **281**, 487

Kauffmann, G.,Charlot, S., White, S.D.M.: 1996, *MNRAS* **283**, L117

Kauffmann, G., Charlot, S.: 1998, *MNRAS* **297**, L23

Kauffmann, G., et al.: 1999, *MNRAS* **303**, 188

Kodama, T., Bower, R.G., Bell, E.F.: 1999, *MNRAS* **306**, 561

Larson, R.B.: 1974, *MNRAS* **173**, 671

Lilly, S.J., et al.: 1996, *ApJ* **460**, L1

Madau, P., Pozzetti, L., Dickinson, M.: 1998, *ApJ* **498**, 106

Menanteau, F., Abraham, R.G., Ellis, R.S.: 2001, *MNRAS* **322**, 1

Menci, N., Cavaliere, A., Fontana, A., Giallongo, E., Poli, F.: 2002, *ApJ* **575**, 18

Miyazaki, M., et al.: 2002, *ApJ* submitted, [astro-ph/0210509]

Mohan, N.R., et al.: 2002, *A&A* **383**, 440

Peebles, P.J.E., 2002: [astro-ph/0201015]

Pozzetti, L., Bruzual, A.G., Zamorani, G.: 1996, *MNRAS* **281**, 953

Pozzetti, L., et al.: 1998, *MNRAS* **298**, 1133

Pozzetti, L., et al.: 2003, *A&A* **402**, 837

Primack, J.R.: 2002, astro-ph/0205391

Renzini, A.: 1999, in *The Formation of Galactic Bulges*, (eds) C.M. Carollo, H.C. Ferguson, & R.F.G. Wyse (Cambridge: CUP), p. 9

Renzini, A., Cimatti, A.: 1999, *A.S.P. Conf. Ser.* **193**, 312

Rodighiero, G., Franceschini, A., Fasano, G.: 2001, *MNRAS* **324**, 491

Roche, N.D., et al.: 2002, *MNRAS* **337**, 1282

Sandage, A., Tammann, G.A., Yahil, A.: 1979, *ApJ* **232**, 352

Schade, D., et al.: 1999, *ApJ* **525**, 31

Schmidt, M.: 1968, *ApJ* **151**, 393

Smail, I., et al.: 2002, *ApJ* **582**, 844

Smith, G.P., et al.: 2002, *MNRAS* **330**, 1

Somerville, R.S, Primack, J.R., Faber, S.M.: 2001,*MNRAS* **320**, 504

Spinrad, H., et al.: 1997, *ApJ* **484**, 581

Stanford, S.A., Eisenhardt, P.R., Dickinson, M.: 1998, *ApJ* **492**,461

Stern, D., Connolly, A., Eisenhardt, P., et al.: 2001, in "Deep Fields", Proceedings of the ESO Workshop, Garching, Germany, Springer-Verlag, p. 76

Stiavelli, M., et al.: 1999, *A&A* **343**, L25

Andrea Cimatti

Tinsley, B.M.: 1972, *ApJ* **178**, 319
Totani, T., Yoshii, Y.: 1998, *ApJ* **501**, L177
Totani, T., Yoshii, Y., Maihara, T., Iwamuro, F., Motohara, K.: 2001, *ApJ* **559**, 592
Treu, T., et al.: 2002, *ApJ* **564**, L13
van Albada, T.S.: 1982, *MNRAS* **201**, 939
Waddington, I., et al.: 2002, *MNRAS* **336**, 1342
Wehner, E.H., Barger, A.J., Kneib, J.-P.: 2002, *ApJ* **577**, L83
White, S.D.M., Rees, M.J.: 1978, *MNRAS* **183**, 341
Zepf, S.E.: 1997, *Nat* **390**, 377

A "Clear" View of the Nucleus: the Megamaser Perspective

Hans–Rainer Klöckner[1,2] and Willem A. Baan[2]

[1] *Kapteyn Institute, University of Groningen, P.O. Box 800, 9700 AV Groningen, The Netherlands*

[2] **ASTRON**, *Westerbork Observatory, P.O. Box 2, 7990 AA Dwingeloo, The Netherlands*

2002 November 21

Abstract. Extragalactic emission from the hydroxyl and the water molecule was first detected in the early eighties, revealing a new class of maser emission with unexpected isotropic luminosities of many magnitudes higher than their galactic counterparts. Galaxies that harbor this so–called Megamaser emission show enhanced core activity in the form of a nuclear starburst or an active–galactic-nucleus. The exceptional maser properties together with the nuclear activity indicate that the line radiation originates in the circumnuclear environment close to the central engine. The environment for producing maser emission in our Galaxy fulfills some unique requirements that will be compared with those of the extra–galactic Megamaser emission. Using very–long–baseline–interferometry, the observational data show that the radio and the molecular line emission structure reveal a rather more complex picture of the circumnuclear environment where the masers occur. At such scale–sizes the individual Megamaser galaxies display diverse maser- and nuclear properties, which all contribute to the understanding of the molecular environment in active nuclei.

Keywords: molecules, maser emission, extra–galactic maser, active galactic nuclei, starburst nuclei

1. Origin of galactic maser emission

The first detections of *mysterious* microwave emission lines were made toward several HII regions, which had known hydroxyl (OH) absorption lines. These lines showed rather unusual properties that could not been explained by the authors at the time (Weaver et al., 1965). Shortly after these observations the lines were interpreted as maser emission, which provided a possible explanation of the emission process, and the extreme brightness temperatures, the polarization, and the line ratios (Perkins et al., 1966). Up till now several molecular species have been found to exhibit maser emission (see Table 1) and have provided crucial information on the maser environment itself and on the dynamics of the local surrounding medium. Maser emission can serve as an accurate tool to determine distances, independent from the known distance ladder based on proper motion (Genzel et al., 1981) or phase lag measurements (Herman et al., 1985), and to determine magnetic field strengths by using the Zeeman effect (Reid & Silverstein, 1990).

Maser emission has been detected in the Galaxy in starformation regions, planetary nebulae, circum–stellar disks, and stellar envelopes, revealing the

M.J.P.F.G. Monteiro (ed.), The Unsolved Universe: Challenges for the Future, 49-57.
© 2003 *Kluwer Academic Publishers.*

Table I. Molecules exhibiting maser emission.
References: [a]Weaver et al. (1965), [b]Cheung et al. (1969),
[c]Barrett et al. (1971), [d]Barrett et al. (1971), [e]Turner &
Zuckerman (1973), [f]Snyder & Buhl (1974), [g]Wilson et al.
(1982), and [h]Guilloteau et al. (1987).

Molecule		Year of publication
Hydroxyl	OH	1965[a]
Water vapor	H_2O	1969[b]
Methanol	CH_3OH	1971[c]
Formaldehyde	H_2CO	1971[d]
Methylidyne radical	CH	1973[e]
Silicon Oxide	SiO	1974[f]
Ammonia	NH_3	1982[g]
Hydrogen cyanide	HCN	1987[h]

physical conditions at scalesizes ranging from astronomical units up to hundreds of parsecs. Most research has been done on the *classical* masering molecules H_2O and OH in the Galaxy as well as in extra–galactic sources. Therefore, the following sections will review the properties of the extra–galactic detections in the context of galactic properties.

The hydroxyl emission has been detected at several rotational levels, where in particular the ground state transitions comprise four lines: the satellite lines (1612 and 1720 MHz) and the main lines (1665 and 1667 MHz). Galactic sources display anomalous line ratios as compared with the LTE line–ratios for the main- and satellite lines of 1–1–5–9. The spectra of the satellite lines and the 1665 MHz main–line observed in starformation regions are complicated by a general lack of spatial resolution. Within a sample of starformation regions several 1612 MHz masers are associated with the late–type stars, whereas the 1720 MHz maser lines are associated with supernova remnants (Caswell, 1999). The OH satellite maser lines in stellar envelopes display weak polarization combined with enhanced 1612 MHz emission. These Type II[1] stars are mostly late–type stars showing moderate to large infra–red excess indicating a thick, dusty envelope. Stars with main line emission (Type I) are grouped predominantly with steeper infra–red colors (Olnon, 1977), and a dominant 1665 MHz lines suggests a larger dust spectral index and a distinct dust composition (Elitzur, 1992b). In contrast to the stellar maser lines, the 1665 MHz OH main–line is pronounced toward HII/OH regions (e.g. the strongest OH emission is found in W3(OH), a prototype HII re-

---

[1] Historically Type I stars display the main lines while Type II stars show the satellite lines.

gion) and displays enhanced and polarized emission in combination with weak 1720 MHz satellite emission. Such maser spots are located on the circumference of the associated HII region and the environment traced by the masers is the post–shock envelope of an expanding and compressed shell (Elitzur & de Jong, 1978). Such masers are part of the short–lived early (10^4 years) phase in the evolution of HII regions.

Although the maser line ratios are quite different in all of these environments, it is most likely that OH is radiatively pumped by the infra–red radiation field (e.g. Elitzur, 1992a). The only exception would be the enhanced 1720 MHz emission line, which has been the least commonly observed and which is an indicator for interactions between supernova remnants and molecular clouds. The current excitation models show that C–shocks (≤ 50 km s^{-1}) can produce OH 1720 MHz maser conditions by collisional pumping (Lockett et al., 1999).

The water vapor masers at 22 GHz has been found in similar sources as the hydroxyl masers, but the emission features are spatially separated from the OH. Model calculations and observations show that the 22 GHz line is not unique and that more transitions can have inverted level populations (e.g. 321 GHz sub–mm maser, Menten et al., 1990). H_2O masers are detected towards many starformation regions and there is an excellent correlation between the occurrence of compact HII regions and OH masers. However, their morphology is quite complex with several regions comprising many maser features but without a clear underlying structure. The maser emission is always separate from and not physically related to the HII/OH regions; they are mostly associated with molecular mass outflows and newly formed stars (Reid & Moran, 1981). W49N is the strongest galactic water vapor emission detected with a total luminosity of around 1 $L_\odot$ and with individual spots of maximally 0.08 $L_\odot$ (Walker et al., 1982).

In the stellar environments H_2O has been frequently detected in association with OH emission as well but in a typically 15–50 times smaller stellar envelope. Here the pumping mechanism is collisional excitation of the rotational levels, which relates the stellar mass loss rate to the maser luminosity (Cooke & Elitzur, 1985). The same pumping mechanism also plays a role in starformation regions with post J–shocks environments.

2. OH and H_2O extra–galactic Megamaser

Galactic masers are found to be signposts of starformation and the detection of the first extra–galactic H_2O and OH maser–emission in the late seven-

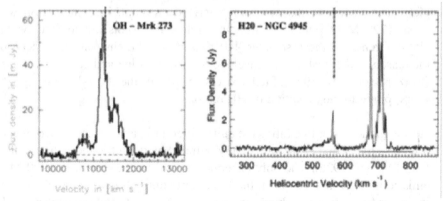

Figure 1. Representative spectra of two individual Megamaser galaxies.
Left: A spectrum of the OH Megamaser emission in Mrk 273 observed with the WSRT interferometer. Right: A spectrum of H_2O maser emission in NGC 4945 observed with the Parkes
radio telescope of the CSIRO (Greenhill et al., 1997). The velocity spread is shown relative to
the systemic velocity of the galaxy indicated by the vertical bar.

ties made it clear that this emission mirrors the known galactic counterparts.
These maser features with slightly higher luminosities were associated with
HII or starformation regions displaced from the galactic centers (e.g. M 82,
NGC 253, M 33). Soon after a new sort of extra–galactic maser emission was
discovered toward the nuclear regions having isotropic luminosities of six or
more magnitudes higher than the galactic sources. The luminosity and the
exceptional line width suggested that this emission might trace the circumnuclear environment and expose the nuclear properties (H_2O Dos Santos et al.,
1979, OH Baan et al., 1982).

Although the luminosity of the H_2O Megamasers (H_2O–MMs) is much higher,
they exhibit similar properties as the galactic H_2O masers (e.g. polarization).
On the other hand, the characteristics of extra–galactic OH Megamasers are
distinctly different from their galactic counterparts; for example, the extra–
galactic 1667 MHz main–line emission is found to be enhanced relatively
to the 1665 MHz line and is unpolarized. The emission spectra of two Megamaser galaxies are shown in Figure 1 in order to display the spectroscopic signatures of both molecular types. The OH spectra show a continuous (smooth)
emission profile with a velocity width of up to thousands of km s^{-1}, sometimes with overlapping OH main–lines. The H_2O Megamasers show distinct
and narrow emission features spread over an equivalent velocity range. Such
distinction is also found for stellar OH & H_2O maser emission. Observational
data suggest that the spectra of H_2O Megamasers is made up of predominantly narrow components (typical line widths of 1.5 km s^{-1}; M 33), while
the spectra of OH Megamasers may result from (many more) broader compo-

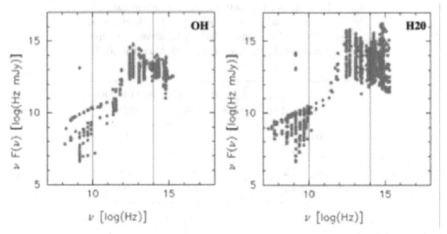

Figure 2. Spectral energy distribution of the integrated OH (left panel) and H_2O Megamaser samples. The emission at each band from every galaxy has been plotted into one frame in order to display the general energy release of the Megamaser galaxies. The frequency range covers seven magnitudes starting in the radio, to the infra–red, up to the visible regime, whereas the vertical lines indicate the boundaries between these regimes. Note that the H_2O Megamasers are found at lower redshifts and have higher fluxes than the OH galaxies.

nents (typical line widths of 10 km s^{-1}; M 82). The picture for H_2O sources is supported by observations of strong H_2O Megamasers showing distinct line features flaring on shorter than dynamical timescales.

The extra–galactic Megamaser sources are morphologically distinct. While the H_2O Megamasers are mostly related to early– to late–type spiral galaxies (Braatz et al., 1997), the OH emitters show peculiar and irregular structures as evidence of merging scenarios. The total energy release in both types is displayed in Figure 2, which shows a slightly different spectral energy distribution (SED) from the radio–to–optical wavelengths. The SED of the OH emitters displays the tightest distribution relating to a rather uniform emission mechanism. A comparison of the optical and infra–red emission components of OH and H_2O sources show that the hydroxyl galaxies have most of their emission in the infra–red bands, whereas the H_2O galaxies emit a similar amount of energy within the optical and infra–red regimes. The energetic signature of OH–MM galaxies mimics a typical starforming region with thermal free–free emission in the radio and enhanced dust emission with temperature of about 60 K in the infra–red. The infra–red luminosity in these sources is exceptionally high making them a sub–sample of the ultra–luminous infra–red galaxies. The extreme infra–red radiation field is thought to provide the radiative pump for the maser emission (Baan et al., 1989). The equal energy release of the H_2O Megamaser galaxies in the optical and infra–red bands and the larger spread in data points may reflect a lower dust content

Table II. Nuclear activity of Megamaser galaxies. A compilation of the optical classification of the OH and H_2O Megamaser sample (Greenhill, 2001, Baan et al., 1998, Maloney, 2002).

classified / detected	Hydroxyl 23/99	Water vapour 16/30
SBN	10	1
LINER	7	5
Seyfert 2	5	10
Seyfert 1	1	0

of these galaxies and a different nuclear power plant. The classification of the dominant nuclear energy source of the Megamasers is presented in Table 2. The general trends seen from the galactic SEDs are in agreement with these classifications, such that the majority of the hydroxyl emitters shows starburst (SBN) phenomena and the water vapor masers are mostly related to Seyfert 2 type nuclei (AGN).

3. Nuclear composition of Megamaser galaxies

A small fraction of the Megamaser galaxies has been observed at high resolution with very–long–baseline–interferometry (seven OH and nine H_2O sources). These observations of the galactic nuclei show a complex picture for the continuum and the maser emission. A comparison of both types can only be done indirectly using the spatial and kinematical properties, because the OH and H_2O masers do not occur in the same galaxies, except for one non–nuclear detection of OH emission in the H_2O–MM galaxy NGC 1068 (Gallimore et al., 1996). Therefore, at parsec resolution the individual H_2O emission features remain unresolved, whereas the hydroxyl maser emission may extend over hundreds of parsec. At VLBI resolutions as much as 50 percent of the OH emission is not detected. The OH masers are infra–red pumped reflecting dust temperatures of around 45 K (Baan et al., 1989; Skinner et al., 1997) and trace a "mutually exclusive" environment with respect to the water–vapor masers. The nuclear H_2O masers are found to be associated mostly with accretion disks, while the masers away from the center are associated with jets; both types show collisional pumping at temperatures of more than 250 K (Kartje et al., 1999).

The OH–MM prototype Arp 220 displays maser emission at each of its two nuclei, which display the characteristics of the orbital dynamics of the nuclei (Baan et al., 1995). The integrated emission spectrum indicates a distinctly non–LTE line ratio. The strongest emission feature is dominated by the 1667 MHz from the western nucleus, while the second feature represents a mix of the 1665 MHz line from the western nucleus and the weaker 1667 MHz line from the eastern nucleus. A third feature results from the 1665 MHz emission from the East. At high resolution the observed maser emission splits into individual bright or diffuse regions surrounding the central continuum emission (Lonsdale et al., 1998). The emission itself shows a rather poor association with the continuum emission, as would be expected for the classical OH Megamaser model (Baan et al., 1989), which raises the question of pump efficiency within the individual emission regions. The more extended and weaker emission regions have a rather broad line profile, which may be correlated with the boundary of the nuclear bubble indicated by Chandra observations (Clements 2002, xxx/0208477 preprint). In combination with the X–ray observations, another more compact OH emission region with a line width of hundreds of km s^{-1} might be connected with a nuclear disk or torus obscuring a weak AGN.

The existence of such larger scale nuclear structures has been suggested by theoretical work and a first example may indeed have been detected in the Seyfert 1 type nucleus in the OH–MM Mrk 231 (Krolik, 1999; Klöckner & Baan, 2001). The Mrk 231 observations show extended OH emission of a hundred parsecs in a half–circular shape straddling the nuclear radio source. The emission reveals a rather smooth environment across the whole region with LTE OH main–line ratios and suggesting optically thin masering (Klöckner et al., in prep.). Modeling suggests an inclined thick disk or TORUS of 200 pc in size, which would provide high enough column densities to account for the observed obscuration and for the maser emission toward an active nucleus. In addition, the OH–MM Mrk 273 shows a similar structure embedded in a starburst nucleus. In this galaxy the OH emission clearly traces the circumnuclear dynamics of a thick disk or TORUS of 180 pc, that is oriented almost perpendicularly to the kinematical structure at larger scales. The distinct velocity pattern at the center of this structure suggests a moderately radio weak AGN with a central binding mass of $3.5 \times 10^8 M_\odot$ (Klöckner & Baan, 2002).

The H_2O–MM emission in the Seyfert 2 galaxy NGC 4258 reflects the most reliable determination of a pc–scale disk signature close to the central engine (Greenhill et al., 1995). The integrated line profile shows a symmetry with respect to the systemic velocity. High–resolution observations accurately measure the individual emission features tracing the Keplerian rotation in the nuclear region. This well–behaved motion constrains the central mass distribution to be a black hole of $\sim 10^7 M_\odot$. It cannot be ruled out whether the maser spots actually reflect a true disk ranging in size from 0.14 to 0.28 pc or that they trace a thin molecular layer within a thicker envelope (Herrnstein et al., 1998). In NGC 1068 the combination of continuum and maser observations reveals a edge–on molecular structure of 1.3–2.5 pc that represents the outer part of an accretion disk at the nucleus (Gallimore et al., 1997). Optical polarimetry of NGC 1068 indeed suggests the presence of a Seyfert 1 nucleus hidden from direct view and visible only in polarized light resulting from dust obscuration (Miller et al., 1991). Maser emission at the highest spatial resolution does not always easily reveal the secrets of the nuclear region as seen in NGC 1068. In the case of the galaxy NGC 3079, the integrated emission spectrum is asymmetric with respect to the systemic velocity with strong blue–shifted lines and red–shifted lines that are two orders of magnitude weaker (Hagiwara et al., 2002). At high angular resolution the individual maser features are distributed both perpendicular and parallel with respect to the nuclear continuum emission. Although the traced velocity pattern does not clearly show any systematic motion, the observed structures may finally help to identify the central of the three continuum components as the location of the nucleus.

Acknowledgements

The European VLBI Network is a joint facility of European and Chinese radio astronomy institutes funded by their national research councils.

The Westerbork Synthesis Radio Telescope is operated by ASTRON (Netherlands Foundation for Research in Astronomy) with support from the Netherlands Foundation for Scientific Research NWO.

This research has made use of the NASA/IPAC Extragalactic Database (NED), which is operated by the Jet Propulsion Laboratory, California Institute of Technology, under contract with the National Aeronautics and Space Administration.

References

Baan, W.A.: 1989, *ApJ* **338**, 804.
Baan, W.A., Haschick, A.D.: 1995, *ApJ* **454**, 745.
Baan, W.A., Salzer, J.J., Lewinter, R.D.: 1998, *ApJ* **509**, 633.
Baan, W.A., Wood, P.A.D., Haschick, A.D.: 1982, *ApJL* **260**, L49.

Barrett, A.H., Schwartz, P.R., Waters, J.W.: 1971, *ApJL* **168**, L101.

Braatz, J.A., Wilson, A.S., Henkel, C.: 1997, *ApJS* **110**, 321.

Caswell, J.L.: 1999, *MNRAS* **308**, 683.

Cheung, A.C., Rank, D.M., Townes, D.D., Thornton, C.H., Welch, W.J.: 1969. *Nat* **221**, 626.

Cooke, B., Elitzur, M.: 1985, *ApJ* **295**, 175.

Dos Santos, P.M., Lepine, J.R.D.: 1979, *Nat* **278**, 34.

Elitzur, M.: 1992, *ARAA* **30**, 75.

Elitzur, M.: 1992a, in *Astronomical masers*. Kluwer Academic Publishers.

Elitzur, M., de Jong, T.: 1978, *A&A* **67**, 323.

Gallimore, J.F., Baum, S.A., O'Dea, C.P.: 1997, *Nat* **388**, 852.

Gallimore, J.F., Baum, S.A., O'Dea, C.P., Brinks, E., Pedlar, A.: 1996, *ApJ* **462**, 740.

Genzel, R., Reid, M.J., Moran, J.M., Downes, D.: 1981, *ApJ* **244**, 884.

Greenhill, L.J., Jiang, D.R., Moran, J.M., Reid, M.J., Lo, K.Y., Claussen, M.J.: 1995, *ApJ* **440**, 619.

Greenhill, L.J., Moran, J.M., Herrnstein, J.R.: 1997, *ApJL* **481**, L23.

Greenhill, L.J.: 2001, In: *IAU Symposium 206: Cosmic Masers: from protostars to blackholes*; (eds) V. Migenes, E. Luedke.

Guilloteau, S., Omont, A., Lucas, R.: 1987, *A&A* **176**, L24.

Hagiwara, Y., Henkel, C., Sherwood, W.A., Baan, W.A.: 2002, *A&A* **387**, L29.

Herman, J., Habing, H.J.: 1985, *A&AS* **59**, 523.

Herrnstein, J.R., Moran, J.M., Greenhill, L.J., Blackman, E.G., Diamond, P.J.: 1998, *ApJ* **508**, 243.

Kartje, J.F., Königl, A., Elitzur, M.: 1999, *ApJ* **513**, 180.

Klöckner, H.-R., Baan, W.A.: 2001, In: *IAU Symposium 206: Cosmic Masers: from protostars to blackholes*; (eds) V. Migenes, E. Luedke.

Klöckner, H.-R., Baan, W.A., Garrett, G.A.: 2002, p. in prep.

Klöckner, H.-R. and W.A. Baan: 2002, In *Proceedings of the 5th European VLBI Network Symposium*, Gustav–Stresemann–Institut, Bonn, Gemany, (eds) E. Ros, R.W. Porcas, A.P. Lobanov, J. A. Zensus, pp. 175.

Krolik, J.H.: 1999, in *Active galactic nuclei: from the central black hole to the galactic environment*, (eds) Julian H.Krolik.Princeton, N.J., Princeton University Press.

Lockett, P., Gauthier, E., Elitzur, M.: 1999, *ApJ* **511**, 235.

Lonsdale, C.J., Lonsdale, C.J., Diamond, P.J., Smith, H.E.: 1998, *ApJL* **493**, L13.

Maloney, P.R.: 2002, *Pub. of the Astron. Soc. of Australia* **19**, 401.

Menten, K.M., Melnick, G.J., Phillips, T.G., Neufeld, D.A.: 1990, *ApJL* **363**, L27.

Miller, J.S., Goodrich, R.W., Mathews, W.G.: 1991, *ApJ* **378**, 47.

Olnon, F.M.: 1977, *Ph.D.Thesis; Sterrewacht Leiden*.

Perkins, F., Gold, T., Salpeter, E.E.: 1966, *ApJ* **145**, 361.

Reid, M.J., Moran, J.M.: 1981, *ARA&A* **19**, 231.

Reid, M.J., Silverstein, E.M.: 1990, *ApJ* **361**, 483.

Skinner, C.J., Smith, H.A., Sturm, E., Barlow, M.J., Cohen, R.J., Stacey, G.J.: 1997, *Nat* **386**, 472.

Snyder, L.E., Buhl, D.: 1974, *ApJL* **189**, L31.

Turner, B.E., Zuckerman, B.: 1973, *BAAS* **5**, 420.

Walker, R.C., Matsakis, D.N., Garcia-Barreto, J.A.: 1982, *ApJ* **255**, 128.

Weaver, H., David, W.W., Dieter, N.H., Lum, W.T.: 1965, *Nat* **208**, 29.

Wilson, T.L., Batria, W., Pauls, T.A.: 1982, *A&A* **110**, L20.

Low Luminosity BL Lacs

Sónia Antón (sonia.anton@oal.ul.pt)
Centro de Astronomia e Astrofísica da Universidade de Lisboa
Observatório Astronómico de Lisboa, Tapada da Ajuda, 1349-018 Lisboa

2002 December 4

Abstract. There are radio sources that have prominent synchrotron cores, and in some, e.g. BL Lacs, the non-thermal emission is still prominent at infrared/optical wavelengths. We have been analysing the Spectral Energy Distributions (SEDs) of a sample of radio-loud flat-spectrum low-luminosity objects that show a remarkable distribution of optical activity: from "Seyfert-like" objects, BL Lac objects to "normal" elliptical galaxies. The aim is to investigate if there is any correlation between differences in the SED and the optical classification. The results indicate that part of the optical diversity is not intrinsic, and that it is correlated with the frequency at which the synchrotron emission begins to decline.

1. Introduction

There are galaxies that radiate large amounts of energy, often over a wide range of frequencies, by processes other than the emission of starlight. The emission comes from the very centre of the galaxy and for this reason we talk of Active Galactic Nuclei (AGN). Observationally this extra activity shows up in a variety of ways, and according to its characteristics, the object is classified into different groups. Considering the emission in the radio domain, AGNs came, broadly, in two flavours: those with weak radio emission, the "radio-quiet", and those with strong radio emission, the "radio-loud". Amongst the radio-loud AGN, we find at high luminosities objects with strong optical emission lines, the Quasars, and at lower luminosities objects with very weak or absent emission lines, the BL Lacs.

2. Some properties of the BL Lacs

BL Lacs are found in elliptical galaxies, showing up in the radio as an unresolved bright core or core-jet source. The broadband spectrum is smooth and flat from the radio up to the infrared/optical bands (or beyond that), the emission is variable sometimes on timescales as short as minutes to hours, and is often polarised (see e.g. Urry & Padovani, 1995; and Kollgard, 1994, for reviews). It is believed that the continuum radiation is dominated by non-thermal emission from a relativistic jet pointing to the observer at a small angle with the line of sight, as proposed by Blandford & Rees (1978). The Spectral Energy Distributions (SED) of BL Lacs are characterized by having

M.J.P.F.G. Monteiro (ed.), The Unsolved Universe: Challenges for the Future, 59-68.

a 2-bump shape, in units of ν F vs. ν (ν is frequency, F the flux density). The first bump is related to synchrotron radiation, and the second bump is related to inverse Compton scattering. The first bump is characterised by the frequency ν_{peak} at which the synchrotron component begins to decline. This frequency is directly related to the maximum in the synchrotron electron energy spectrum in the jet. Some BL Lacs (and Flat Spectrum Radio Quasars) show the first energy peak in the infrared/optical wavelengths and they are called low-frequency peaked BL Lacs (LBL). On the other hand there are those that show the first energy peak in the EUV region, the high-frequency peaked BL Lacs (HBL). For some time it was thought that BL Lacs belonged to a bimodal LBL-HBL population, but the apparent dichotomy was due to selection effects (see e.g. Padovani, 1999, and references therein). It is now widely accepted that BL Lacs have a broad ν_{peak} distribution, spanning $\sim$ 6 orders of magnitude in frequency ($10^{12} < \nu_{peak} < 10^{18}$ Hz), the extremes being populated by LBLs and HBLs. Fossati et al (1998) and Ghisellini et al. (1998) found a sequence in the frequency at which the peak of the non-thermal emission occurs: the synchrotron peak position and the radio luminosity are anti-correlated in the sense that the lower the peak position the more luminous the source is.

3. BL Lac classification

Apart from the radio-loudness, core-jet morphology and flatness of the broadband spectrum, an object is classified as a BL Lac if it has weak emission lines and a strong non-thermal optical component relative to the starlight. Those are commonly quantified by the equivalent width (EW) of the brightest emission line, and the strength of the non-thermal optical component measured through the relative depression of the continuum at 4000 Å, the 4000 Å break contrast C[1]. The classical definition classifies as BL Lacs those objects that have EW< 5Å and C< 0.25 (e.g. Stocke et al., 1991). These are empirical limits, based on the properties of a relatively small number of objects. Results from newer samples have been showing that in a C-EW plot, neither BL Lacs and other "non-BL Lac" objects are clearly separated (Marchã et al., 1996; hereafter M96), nor the 4000Å break contrast distribution of BL Lacs and other "non-BL Lac" objects reveal a clear bimodality (Laurent-Muehleisen et al., 1998; Caccianiga et al., 1999). As first noted by M96 the traditional classification of

---

[1] The spectrum of a normal galaxy shows a flux discontinuity (CaII H & K break) at 4000 Å rest, which is quantified by the difference between the redward and the blueward flux densities of the break – the contrast C. Early-type galaxies show a very narrow range of contrasts, C=0.49±0.1. If it happens that besides the thermal emission there is an extra continuum, then that depression will be decreased by some amount. Note that the stronger the extra component is the smaller C is.

Table I. 200 mJy sample selection criteria. The spectral index α is defined between 1.4 GHz (NVSS catalogue) and 5 GHz (GB6 catalogue), R magnitudes are from APM, corrected by the calibration presented in Antòn (2000).

S_{5GHz}	$\alpha_{1.4-5GHz}$	R	z	δ_{1950}	$|b|$
≥ 200 mJy	≥ -0.5	≤ 17 mag	≤ 0.2	$\geq 20^o$	$\geq 12^o$

a BL Lac is too restrictive, and an "expanded" definition that includes objects with $C < 0.4$ has been proposed.

4. The 200 mJy sample

The research on BL Lacs was first based on very bright and relatively small samples, either selected on the radio band, e.g. "1 Jy sample" (Stickel et al, 1991) or in the X-ray band, e.g. EMSS[2] (Gioia et al., 1990; Stocke et al., 1991; Maccacaro et al., 1994). As a consequence, the knowledge on BL Lacs has been built on the properties of the most powerful objects, which may not be representative of the whole population. In fact, the results from the newer, deeper and larger samples, e.g. REX[3] (Caccianiga et al., 1999), RGB[4] (Laurent-Muehleisen et al., 1998) and DXRBS[5] (Perlman et al., 1998), have been revealing BL Lacs with intermediate properties between the objects found in bright radio-selected samples and in bright X-ray selected samples. The 200 mJy sample is a relatively low-luminosity radio-selected sample, that comprises all the JVAS[6] (Patnaik et al., 1992; Browne et al., 1998; Wilkinson et al., 1998) sources with core-dominated morphology at 8.4 GHz VLA in A configuration and obey the criteria presented in Table I. The extended radio emission is below the FRI/FRII division (Dennett-Thorpe & Marchã, 2000). Thus, the radio properties of the 200 mJy objects are similar to those found in BL Lacs. M96 obtained optical spectra and optical polarization for a large number of 200 mJy objects. They found that the homogeneity disappears at the optical wavelengths: 30% of the objects are BL Lacs but the remaining 70% show a range of optical activity. They classify the objects in different optical classes, which in terms of the EW and 4000Å break contrast can be summarised as:

[2] Einstein Extended Medium Sensitivity Survey
[3] Radio-Emitting X-ray Survey
[4] Rosat All Sky Survey Green Bank
[5] Deep X-ray Radio Blazar Survey
[6] Jodrell-Bank VLA Astrometric Survey

- **BL Lac objects (BLL)**. These are featureless spectra objects which obey the "classical" BL Lac definition: $C < 0.25$ and $EW < 5$ Å.

- **BL Lac Candidates (BLC)**. These are objects with break contrast of $C < 0.35$ and $EW < 40$ Å. Even if both quantities are larger than the "classical" definition, the 4000 Å break contrast suggests the existence of an extra-continuum, and regarding the small EW of the emission lines, the extra-continuum is unlikely to be thermal.

- **Seyfert-like objects**. These are the objects with $EW > 40$ Å, comprising broad emission line objects and narrow emission line objects.

- **Passive Elliptical Galaxies (PEGs)**. These are objects with break contrast > 0.39 and weak emission lines $EW < 30$-40 Å.

We have been analysing the 200 mJy sample with the aim of identifying the origin of the diversity. In particular, we embarked in a multiwavelength study to investigate if there was any correlation between differences in the SED and the optical classification. One of main goals was to establish how far the synchrotron component extends in frequency and to estimate the frequency at which the break in the non-thermal spectrum of those objects that are not obvious BL Lacs occurs. For this study, new data was gathered: submillimetre data with SCUBA, infrared data with ISO and optical data with NOT[7]. As a result, the SEDs of the objects were produced; some examples are presented in Figures 1 and 2.

5. The Spectral Energy Distributions

The analysis of the SEDs allowed us to conclude that the objects split mainly in two broad groups:

- One third of the objects show convex broadband spectra between the radio and the millimetre bands, the non-thermal break occurring between the centimetre and millimetre wavelengths; an example is presented in Figure 1. VLBI observations are available in the literature for some of the objects. For them, the radio images show that the morphology is similar to that of Compact Symmetric Objects (CSOs). This group of convex broadband spectrum objects also includes well known Gigahertz Peaked Sources and starburst galaxies. The spectral and structural information indicate that the nature of these objects might be different from core-jet nuclei, therefore they are not *bona fide* BL Lac candidates.

---

[7] Submillimetre Common-User Bolometer Array, Infrared Space Observatory and Nordic Optical Telescope.

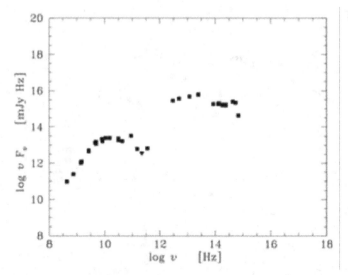

Figure 1. The Spectral Energy distribution, in units of log(ν F) vs. log(ν), of the object 1404+286. Its broadband spectrum between the radio and infrared bands is convex, the non-thermal emission begins to decline at frequencies of few GHz. Upper limits are represented by inverted triangles.

- The remaining sources show smooth and flat (F∼ ν^{α}, $\alpha > -0.5$) broadband spectra up to the millimetre/sub-millimetre wavelengths, sometimes even up to higher frequencies (see Figure 2). When observed with VLBI techniques all the objects show core-jet morphology. In these objects, both spectral and morphological information suggest that the emission is synchrotron from the nuclear core-jet. According to their optical properties the objects are divided in the following categories: 44% objects are classified as BL Lacs, 21% objects are classified as BL Lac Candidates, 27% objects are classified as PEGs, and 9% objects are classified as Seyfert-like objects.

5.1. PEAK FREQUENCY DISTRIBUTION

The SEDs of the flat broadband spectrum objects were further analysed and the synchrotron peak frequencies were estimated; the respective distribution can be found in Figure 3. This Figure shows that the ν_{peak} are widely and fairly continuously distributed from 10^{11} to 10^{15} Hz. No clear separation between well known BL Lacs (C < 0.25 & EW < 5 Å) and other objects (C > 0.25 & EW < 60 Å) is found. The ν_{peak} distribution suggests that the objects are drawn from the same population (otherwise one would not expect a continuous distribution), one that contains a wide range of non-thermal cut-offs.

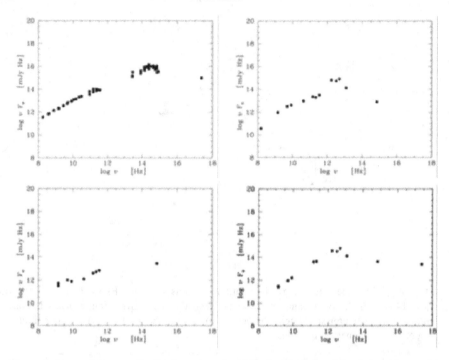

Figure 2. Spectral Energy distributions, in units of log(ν F) vs. log (ν), of the objects with
a flat broadband spectrum. The objects are representative of the different optical classes: the
BL Lac 1652+398 (top left), the PEG 1144+352 (top right), the BL Lac Candidate 1645+292
(bottom left), the Seyfert-like object 1646+499 (bottom right). Upper limits are represented
by inverted triangles.

6. Optical classification and peak frequency distribution

The fact that there is such a wide distribution of ν_{peak} must be relevant for
the classification of an object as a BL Lac. In Figure 4 the values of ν_{peak}
are plotted against the 4000 Å break contrasts. That figure shows that the C
and ν_{cutoff} are correlated: the objects with larger C (PEGs and BLCs) have
lower frequency cut-offs whereas objects that have smaller C (BL Lacs) have
higher frequency cut-offs. Figure 4 shows the obvious: the strength of the
synchrotron component in the optical band is dependent on the frequency
at which the synchrotron emission begins to decline. Note that this fact is
not very important in the case of powerful objects, as in general their non-
thermal optical component is strong enough to fill the 4000 Å break, even
in the case of a low ν_{peak}. But this issue can not be neglected in the case
of low luminosity objects: the correlation shows that the classification of a
low-luminosity object as BL Lac (C < 0.25) is heavily dependent on the
non-thermal cut-off frequency.

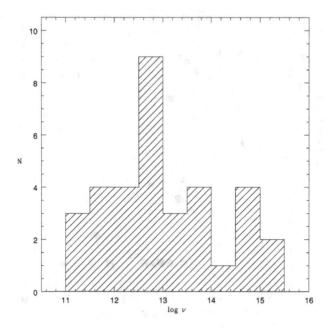

Figure 3. Distribution of the peak frequencies, in log units, of the objects showing smooth and flat broadband spectra. This group includes BL Lacs, BL Lac Candidates, PEGs, and Seyfert-like objects

7. Discussion

We identified a group of objects showing a range of optical activity but with flat broadband spectra between radio and millimetre/sub-millimitre bands, or even beyond. In the framework of unification, both spectral and structural information indicate that these objects have similar properties to those found in the FRI/BL Lac population: the objects discussed here are low-luminosity radio-loud core-dominated objects, their synchrotron emission extends to high frequencies, their extended radio emission is below the FR II range. Optical imaging shows that the objects are hosted by similar elliptical galaxies which live in similar environments (Antón, 2000). More, we found a correlation between the peak frequency and the 4000 Å break contrast, which highlights that the optical classification is dependent on the location of the cut-off of the non-thermal component.

All the evidence points to these objects being similar to BL Lacs, the only difference from object to object is that the first energy peak happens at different frequencies, and for this reason they get classified as different types of sources. In particular, we note that all "classical" BL Lac objects have

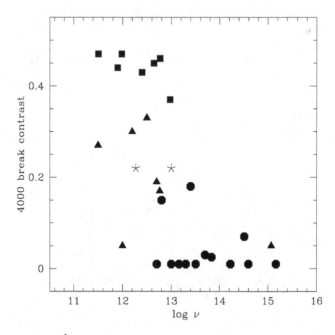

Figure 4. The 4000 Å break contrast vs log ν_{peak} for the objects that have flat broadband spectrum. BL Lacs are represented by circles, BL Lac Candidates are represented by squares, Seyfert-like objects are represented by stars and PEGs are represented by triangles.

$\nu_{peak} > 10^{12.5}$ Hz. We interpret the objects with low ν_{peak} as Very Low frequency Peaked BL Lacs, VLBLs. That is the population of BL Lacs that corresponds to the low-frequency tail of a broad distribution, also comprising frequencies from the infrared/optical bands to the EUV bands. Note that, the existence of a population of very high peaked BL Lacs, BL Lacs that peak at keV frequencies, has been recently proposed ("extreme" BL Lacs; Costamante et al., 2001).

8. Conclusions

We have been studying a sample of low-luminosity radio-loud flat spectrum objects that show a range of activity. Through a multiwavelength study the Spectral Energy Distribution (SED) of these sources were investigated. The main goal was to search for any correlation between differences in the SED and the optical classification. Two types of objects were found: objects with convex spectra between radio and millimetre/submillimetre bands, and objects with a flat spectrum between those bands. Further analysis of the SEDs of the latter group suggest that they are drawn from a single population, but one that has a very broad distribution in non-thermal cut-off frequencies. The spread of ν_{peak} accounts for the observed spread of the strengths of the optical non-thermal emission, and therefore for different optical-classifications.

We conclude that the recognition and classification of a low-luminosity core-jet flat-spectrum object as a BL Lac is heavily dependent on the location of the synchrotron peak frequency. Our results suggest that the classical definition of BL Lacs (EW$\leq$5Å & C$<$ 0.25) is biased toward objects with synchrotron peak frequencies of $\nu_{peak} \geq 10^{12.5}$ Hz. In light of our results, the Very Low Frequency Peaked objects are interpreted as the lower frequency tail version of classical BL Lacs.

Acknowledgements

SA acknowledges financial support from Fundação para a Ciência e a Tecnologia, through grant SFRH/BPD/5692/2001 and parcial funding to attend the JENAM meeting from the JENAM 2002 "Galaxy Evolution" Workshop organizers, through project ref. ESO/PRO/15130/1999 from FCT, Portugal.

References

Antón, S.: 2000, *Ph.D. Thesis*, University of Manchester

Blandford, R.D., Rees, M.J., 1978: in *Pittsburgh Conference on BL Lac Objects*, p. 328

Browne, I.W.A., Marchã, M.J.M.: 1993, *MNRAS* **261**, 795

Browne, I.W.A., Wilkinson, P.N., Patnaik, A.R., Wrobel, J.M.: 1998, *MNRAS* **293**, 257

Caccianiga, A., Maccacaro, T., Wolter, A., della Ceca, R., Gioia, I.M.: 1999, *ApJ* **513**, 51

Costamante, L., Ghisellini, G., Giommi, P., Tagliaferri, G., Celotti, A., Chiaberge, M., Fossati, G., Maraschi, L., Tavecchio, F., Treves, A., Wolter, A.: 2001, *A&A* **371**, 512

Dennett-Thorpe, J., Marchã, M.J.: 2000, *A&A* **361**, 480

Fossati, G., Maraschi, L., Celotti, A., Comastri, A., Ghisellini, G.: 1998, *MNRAS* **299**, 433

Ghisellini, G., Celotti, A., Fossati, G., Maraschi, L., Comastri, A.: 1998, *MNRAS* **301**, 451

Gioia, I.M., Maccacaro, T., Schild, R.E., Wolter, A., Stocke, J.T., Morris, S.L., Henry, J.P.: 1990, *ApJS* **72**, 567

Kollgard, R.I.: 1994, *Vistas in Astronomy* **38**, 29

Laurent-Muehleisen, S.A., Kollgaard, R.I., Ciardullo, R., Feigelson, E.D., Brinkmann, W., Siebert, J.: 1998, *ApJS* **118**, 127

Maccacaro, T., Wolter, A., McLean, B., Gioia, I.M., Stocke, J.T., della Ceca, R., Burg, R., Faccini, R.: 1994, *ApL&C* **29**, 267

Marchã, M.J., Browne, I.W.A., Impey, C.D., Smith, P.S.: 1996, *MNRAS* **281**, 425

Padovani, P.: 1999, in *BL Lac Phenomenon*, *A.S.P. Conf. Ser.* **159**, p. 339

Patnaik, A.R., Browne, I.W.A., Wilkinson, P.N., Wrobel, J.M.: 1992, *MNRAS* **254**, 655

Perlman, E.S., Padovani, P. Giommi, P., Sambruna, R., Jones, L.R., Tzioumis, A., Reynolds, J.: 1998, *AJ* **115**, 1253

Stickel, M., Fried, J.W., Kuhr, H., Padovani, P., Urry, C.M.: 1991, *ApJ* **374**, 431

Stocke, J.T., Morris, S.L., Gioia, I.M., Maccacaro, T., Schild, R., Wolter, A., Fleming, T.A.,
  Henry, J.P.: 1991, *ApJS* **76**, 813
Urry, C.M., Padovani, P.: 1995, *PASP* **107**, 803
Wilkinson, P.N., Browne, I.W.A., Patnaik, A.R., Wrobel, J.M., Sorathia, B.: 1998, *MNRAS*
  300, 790

Cosmic Structures, Parameters & Temperature Anisotropies: Status and Perspectives

François R. Bouchet
Institut d'Astrophysique de Paris, CNRS
98 bis Boulevard Arago, F-75014, France

2003 March 16

Abstract. After recalling the current understanding of the formation of the large scale structures of the Universe which the distribution of galaxies revealed, I review what the imprint on the Cosmic Microwave Background (CMB) of the seeds of these structures can tell us, has already told us, and what it should tell us within the next five years.

Keywords: Cosmology, CMB, Large Scale Structures, Cosmological Parameters

1. The current cosmological paradigm

The analysis of the Cosmic Microwave Background (CMB) temperature anisotropies indicate that we live in a spatially flat Universe, and thus that the total energy density is quite close to the so-called critical density, ρ_c, or equivalently $\Omega = \rho/\rho_c \simeq 1$. About 1/4 of that density appear to be contributed by dark matter - i.e. not interacting electromagnetically - which is cold - i.e. whose velocity dispersion can be neglected) - with $\Omega_{CDM} \simeq 1/4$. If present, a hot dark matter component does not play a significant role in determining the global evolution of the Universe. While many candidate particles have been proposed for this CDM, it has not yet been detected in laboratory experiments, although the sensitivities of the latter are now reaching the range where realistic candidates may lay. The other $\sim$3/4 of the critical density is contributed by a smoothly distributed vacuum energy density or dark energy, whose net effect is repulsive, i.e. it tends to accelerate the expansion of the Universe. Alternatively, this effect might arise from the presence in Einstein's equation of the famous Λ term. The usual atoms (the baryons) in this surprising picture contribute less than about 5 %, i.e. $\Omega_B \simeq 0.05\dots$

The spatial distribution of galaxies revealed the existence of large scale structures (clusters of size $\sim$ 5 Mpc, filaments connecting them, and voids of size $\sim$ 50 Mpc), whose existence and statistical properties can be accounted for by the development of primordial fluctuations by gravitational instability. The current paradigm is that these fluctuations were generated in the very early Universe, probably during an inflationary period; that they evolved linearly during a long period, and more recently reached density contrasts high enough to form bound objects. Given the census given above, the dominant component that can cluster gravitationally is cold dark matter.

M.J.P.F.G. Monteiro (ed.), The Unsolved Universe: Challenges for the Future, 69-86.
© 2003 *Kluwer Academic Publishers.*

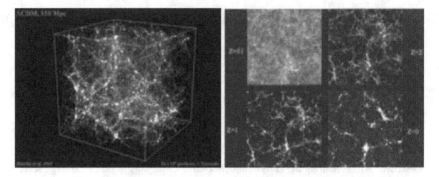

Figure 1. A numerical simulation in a 150 Mpc box of a LCDM Universe ($\Omega = 1$, $\Omega_\Lambda = 2/3$, $n_S = 1$). a) (left) Resulting distribution of the CDM at present (luminosity proportional to the density). b) (right) Temporal evolution by gravitational instability of a thin (15 Mpc) slice of the box showing the hierarchical development of structures within a global cosmic web.

The analysis of the CMB anisotropies also indicate that the initial fluctuations statistics had no large deviations from a Gaussian distribution and that they where mostly adiabatic, i.e. all components (CDM, baryons, photons) had the same spatial distribution. The power spectrum of the initial conditions appears to be closely approximated by a power law of logarithmic slope $n_S = 1$, $P(k) = < \delta_k > \propto k^{n_S}$, where δ_k stands for the Fourier transform of the density contrast ($P(k)$ is therefore the Fourier transform of the two-point spatial correlation function). This shape implies that small scales collapsed first, followed by larger scales, with small objects merging to form bigger objects. The formation of structures thus appear to proceed *hierarchically* within a "cosmic web" of larger structures of increasing contrast.

Figure 1.a shows the generated structures in the CDM components in a numerical simulation box of 150 Mpc, while 1.b shows the evolution with redshift of the density in a thin slice of that box. The statistical properties of the derived distribution (with the cosmological parameters given above) appear to provide a close match to those derived from large galaxy surveys.

When collapsed objects are formed, the baryonic gas initially follows the infall. But shocks will heat that gas, which can later settle in a disk and cool, and form stars and black holes which can then feed back through ionising photons, winds, supernovae... on the evolution of the remaining gas. In this picture, galaxies are therefore (possibly biased) tracers of the underlying large scale structures of the dark matter. This picture has had considerable success, and a few shortfalls which concentrate much current activity...

2. Origin of CMB fluctuations

The CMB is isotropic to a high degree. While in the very early stages of the Universe, radiation and ionised matter where tightly coupled, the Universe became transparent at a redshift of 1100 when the temperature had dropped enough that hydrogen atoms could be stable. This defines a Last Scattering Sphere (LSS) around us which is the ultimate horizon bounding the part of spacetime that we can observe with photons (just like we cannot see further inside that the sun's photosphere). The evolution of primordial fluctuations can be accurately followed and it was long predicted that to account for the formation of large scale structure their imprint as temperature fluctuations should have an rms of $\sim 100\mu K$, which is indeed why it took so long to detect them.

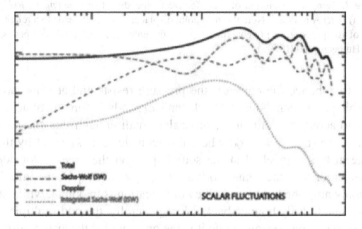

Figure 2. Relative contributions to the angular power spectrum of the temperature anisotropies. It has been assumed that only scalar fluctuations are present.

To analyse the statistical properties of the temperature anisotropies on the LSS, we can either compute the angular correlation function of the temperature contrast δ_T, or the angular power spectrum $C(\ell)$ which is it's spherical harmonics transform (in practice, one transform the δ_T pattern in $a(\ell,m)$ modes and sums over m at each multipole since the pattern should be isotropic, at least for the trivial topology). Figure 2 shows the expected $C(\ell)$ shape in the context we have described above.

This specific shape (for scalar fluctuations) arises from the interplay of several phenomena. One is the so-called "Sachs-Wolf effect" which is the energy loss of photons which must "climb out" of potential well (to ultimately reach us to be observed), an effect which superimposes to the intrinsic temperature fluctuations. Figure 3 shows the (approximate) temporal evolution of the amplitude of modes at different spatial scales (Fourier transform of this intrinsic temperature + Newtonian potential fluctuations field). While gravity

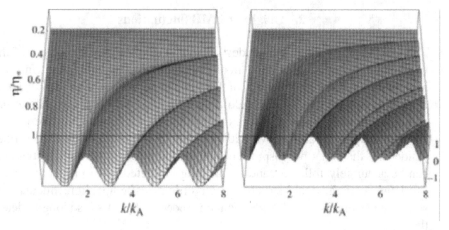

Figure 3. Temporal evolution of the effective temperature (for $R = (p_B + \rho_B)/(p_\gamma + \rho_\gamma) =$ cste). a) (left) Amplitude. Note the zero point displacement which leads to a relative enhancement of compressions. b) (right) rms showing the enhanced odd-numbered peaks. Reprinted from Hu (astroph/0210696).

tends to enhance the contrast, the pressure resists and at some points stops the collapse which bounces back, and expands before recollapsing... This leads to acoustic oscillations, on scales small enough that the pressure can be effective, i.e. for $k > k_A$, where the acoustic scale k_A is set by the inverse of the distance travelled at the sound speed at the time η_* considered. On scales larger than the sound horizon ($k < k_A \propto 1/(c_S \eta_*)$), the initial contrast is simply amplified. At $k = k_A$ the amplification is maximal, while at $k = 2k_A$ it had time to fully bounce back. More generally, the odd-multiples of k_A are at maximal compression, while it is the opposite for the even multiples of k_A.

One should note the displacement of the zero point of the oscillations which results from the added inertia that baryons bring to the CDM+γ fluid. The rms of the modes amplitude (right plot) therefore show a relative enhancement of the odd peaks versus the even ones, this enhancement being proportional to the quantity of baryons, i.e. Ω_B. Let us assume that the LSS transition from opaque to transparent is instantaneous, at $\eta = \eta_*$. What we would see then would just be the direct image of these fluctuations on the LSS; one therefore expects a series of peak at multipoles $\ell_A = k_A \times D_*$, where D_* is the angular distance to LSS which depends on the geometry of the spacetime. Altogether this leads to a dependence of this acoustic scale on the values of cosmological parameters. One finds for instance

$$\frac{\Delta \ell_A}{\ell_A} \simeq -1.1 \frac{\Omega}{\Omega} - 0.24 \frac{\Delta \Omega_M h^2}{\Omega_M h^2} + 0.07 \frac{\Delta \Omega_B h^2}{\Omega_B h^2} \tag{1}$$

around a flat model $\Omega = 1$ with 15% of matter ($\Omega_M h^2 = 0.15$) and 2% of baryons (see Hu astroph/0210696).

Photons must travel through the developing large scale structures to reach the observer. They can lose energy by having to climb out of potential well which are deeper than when they fell in (depending on the rate of growth of structures, which depends in turn on the cosmological census). Of course the reciprocal is also true, i.e. they can gain energy from forming voids. These tend to cancel at small scale since the observer only sees the integrated effect along the line of sight. The red dotted line of Figure 2 shows the typical shape of that Integrated Sachs-Wolf (ISW) contribution. The ISW is anti-correlated with the SW effect, so that the total power spectrum $C(\ell)$ is in fact a bit smaller than the spectrum of each taken separately.

Since the fluid is oscillating, there is also a Doppler effect in the k direction (blue dotted line in Figure 2, which is zero at the acoustic peaks and maximal in between. This effect adds in quadrature (imagine an acoustic wave with k perpendicular to the line of sight, there is no Doppler effect, while for a k parallel to the line of sight the Doppler effect is maximal while the Sachs-Wolf effect is null) which partially smoothes out the peak and trough structure.

So far we considered the fluid a perfect and the transition to transparence as instantaneous, none of which is exactly true. Photons scattered by electrons through Thomson scattering in the baryons-photons fluid perform a random walk and diffuse away proportionally to the square root of time (in comoving coordinates which remove the effect of expansion). Being much more numerous than the electrons by a factor of a billion, they drag the electrons with them (which by collisions drag in turn the protons). Therefore all fluctuations smaller than the diffusion scale are severely dampened. This so-called Silk damping is enhanced by the rapid increase of this diffusion scale during the rapid but not instantaneous combination of electrons and protons which leads to the transparence. As a result (of the finite thickness of the LSS and the imperfection of the fluid), there is an exponential cut-off of the large-ℓ (small scale, $\sim 5'$) part of the angular power spectrum. Finally, other small secondary fluctuations might also leave their imprint, like the lensing of the LSS pattern by the intervening structures, which smoothes slightly the spectrum. But that can be predicted accurately too.

In summary, the seeds of large scale structures must have left an imprint on the CMB, and the characteristics of that imprint can be precisely predicted as a function of the characteristics of the primordial fluctuations and of the homogeneous Universe. Reciprocally, we can use measurements of the temperature anisotropies to constrain those characteristics.

3. A brief historical overview and current status

The first clear detection of the CMB anisotropies was made in 1992 by the DMR experiment aboard COBE (and soon afterwards by FIRS), with a ten

degree (effective) beam and a signal to noise per pixel around 1. This lead
to a clear detection of the large scale, low-ℓ, Sachs-Wolf effect, the flatness
of the curve (see Figure 4.a) indicating that the logarithmic slope of the primordial power spectrum, n_S, could not be far from one. The $\sim 30\mu K$ height
of the plateau gave a direct estimate of the normalisation of the spectrum, A_S
(assuming the simplest theoretical framework, without much possible direct
checks of the other predictions given the data)

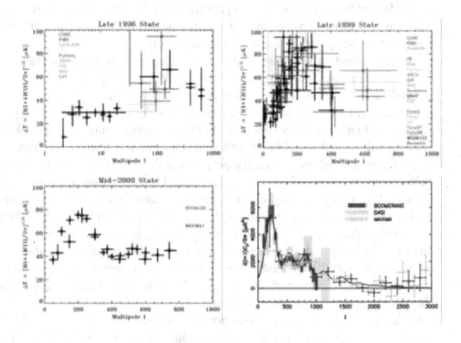

Figure 4. History at a glance through the progressive uncovering of the shape of the power
spectrum of the temperature anisotropies. Top row: the first plot shows all published detection
at the end of 1996, while the second plot is an update at the end of 1999. Bottom row: the
left plot shows the results published in may 2000 by the BOOMERanG and MAXIMA teams
(with each curve moved by + or - 1 σ of their respective calibration, see (Hanany et al., 2000)),
and the right plot (reprinted from (Sievers et al., 2002) compares the CBI results to previous
measurements.

In the next four years (fig 4.a), a number of experiments started to suggest
an increase of power around the degree scale, i.e. at $\ell \sim 200$. As shown
by fig 4.b, by 1999 there was clear indication by many experiments taken
together that a first peak had been detected. But neither the height nor the location of that peak could be determined precisely, in particular in view of the
relative calibration uncertainties (and possible residual systematics errors).

That situation changed in may 2000 when the BOOMERanG and Maxima collaborations both announced a rather precise detection of the power
spectrum from $\ell \sim 50$ to $\ell \gtrsim 600$. That brought a clear determination of the

first peak around an ℓ of 220 (see panel c), with the immediate implications that Ω had to be close to one. This result had considerable resonance since it clearly indicated, after decades of intensive work, that the spatial geometry of the Universe is close to flat, with of course the imprecision due to the poor determination of the other parameters which also have an influence, albeit weaker, on the position of that peak (see eq. 1).

As recalled earlier, a crucial prediction of the simplest adiabatic scenario is the existence of a series of acoustic peaks whose relative contrast between the odd and even ones gives a rather direct handle on the baryonic abundance. In addition, one expects to see at larger ℓ the damping tail. All of these have now been established by the DASI 2001 experiment, an improved analysis of BOOMERanG, and by the release in may of this year of the VSA and CBI results (see panel d). As a cherry on the cake, the first detection of the polarisation of the CMB anisotropies was announced this year too by DASI, at a level consistent with expectations for the emerging model.

The Figure 5 shows the constraints posed by these CMB experiments on some of the parameters of the model. The left column show the result using only weak priors arising from other cosmological studies. It states that the current Hubble "constant", $H_0 = 100h^{-1}$km/s/Mpc, has to have a value between 45 and 90 km/s/Mpc, that the age of the Universe has to be greater than 10 Billion years and that the matter density is larger than 1/10 of the critical density, all of which can be considered as very well established (if for instance the Universe has to be older than it's oldest stars!). The right panel shows the impact of imposing further constraints (on the current amplitude of fluctuations and on the shape of the matter power spectrum) which arise from studies of large scale structures. While quite reasonable, they are some-what less "safe", but illustrate well how the current degeneracies between parameters from CMB alone can be lifted using other probes.

The top panel shows that indeed $\Omega_k = 1 - \Omega$ has to be close to zero. The lower panel shows that Ω_Λ and $\omega_c = \Omega_{CDM}h^2$ are not well determined independently of each other. This simply reflects the fact that the $C(\ell)$ global pattern scales by the angular distance (recall $\ell_A = k_A D_\star$) which is determined by the geometry (i.e. $\Omega = \Omega_\Lambda + \Omega_{CDM} (+\Omega_B)$), while the data is not yet precise enough to uncover the subtler effects which break that degeneracy. It is interesting to recall that the supernovae result is rather sensitive to the quasi orthogonal combination $\Omega_\Lambda - \Omega_{CDM}$. This provides $\Omega_\Lambda \simeq 0.7$, $\Omega_{CDM} \simeq 0.3$ (and $\omega_c \simeq 0.15$ for $h \simeq 0.7$ as suggested by the HST key project), in good agreement with the determination of the CMB using a stronger LSS prior (right panel).

Completing the census, the lowest panel shows the contours in the ω_b - ω_c plane. The CMB determination turns out to be in excellent agreement with the constraints from primordial nucleosynthesis calculations which yield $\omega_b = \Omega_B h^2 = 0.019 \pm 0.002$. Finally, the third panel from the top lends support to

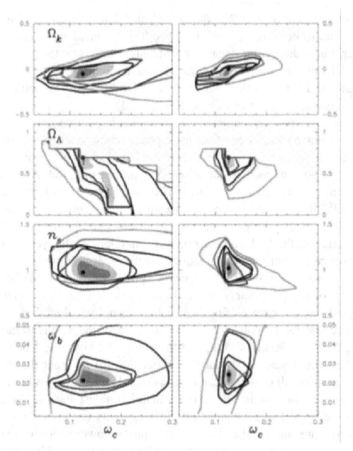

Figure 5. Two-σ likelihood contour plots for some cosmological parameters, obtained by combining DMR with various combination of experiments (with respectively black, purple, dark blue, green and turquoise corresponding to CBI, BOOMERANG, DASI, MAXIMA, and a combination of Boomerang North America + TOCO & all 17 others prior to April 1999). The filled contours correspond to all experiments taken together at the one and two σ level. All panels were made for the "weak-h" prior (i.e. $0.45 < h < 0.9$ & $t_0 > 10$ GYr & $\Omega_m > 0.1$), while the n_s & Ω_b panels add $\Omega_k = 0$. Note that the hatched regions were not searched. Finally, the left panels additionally include a constraint on σ_8 and Γ_{eff} which comes form Large Scale Structures studies. These plots reprinted from (Sievers et al., 2002) illustrate the great consistency achieved between various probes, and the current level of accuracy.

the $n_s = 1$ hypothesis. Many inflationary models suggest value of n_s slightly lower than one (and even departures from a pure power law), but the data is not yet good enough to address these questions.

In summary, this shows that many of the theoretical predictions corresponding to the simplest scenario for the generation of initial conditions (Gaussian statistics, adiabatic modes, no tensorial contribution, a scale invariant power spectrum) in a flat Universe dominated by dark energy and cold dark matter have now been detected, from the Sachs-Wolf plateau, to the series of peaks starting at $\ell \simeq 220$, to the damping tail, together with a first detection of the CMB polarisation at the expected level. The derived parameters are consistent with the various constraints from other cosmological probes and there are no glaring signs of inconsistencies.

This is quite an achievement already, although the CMB constraining power on the theoretical parameters is far from exhausted and a number of crucial predictions of the theory still remain to be checked. Many ground and balloons will undoubtly continue unveiling new aspects of the anisotropies, and in particular in the very short term the Archeops experiment which will improve the determination of the large scale part of the spectrum to further constraint Ω_k (and the other parameters in conjunction with the other one described above).

Major steps should come from the second generation CMB satellite MAP (which has been launched by NASA in June 2001 and whose results should be announced early in 2003) and the third generation one, Planck, to be launched by ESA in early 2007. I now turn to what to expect from these satellites.

4. The future CMB Satellites: MAP & Planck

Both will map the full sky, from an orbit around the Lagrangian point L2 of the Sun-Earth system, to minimise parasitic radiation from Earth. Both are based on the use of off-axis Gregorian telescopes in the 1.5m class. And very importantly for CMB experiments, both will do highly redundant measurements to better detect and remove (or constrain residuals of) possible systematics effect, thanks to the long duration of the data taking (at least a year, to be compared with at most about 10 days for ground experiments which have to cope in addition with the effect of a changing atmosphere - like ozone clouds, the closeness to earth, etc...)

MAP has been designed for rapid implementation, and is based on fully demonstrated solutions. It's observational strategy uses a differential scheme. Two telescopes are put back to back and feed differential radiometers. These radiometers use High Electronic Mobility Transistors (HEMTs) for direct amplification of the radio-frequency (RF) signal. Angular resolutions are not better than 10 minutes of arc.

PLANCK is a more ambitious and complex project, which is designed to be the ultimate experiment in several aspects. In particular, several channels of the High Frequency Instrument (HFI) will reach the ultimate possible sensitivity, limited by the photon noise of the CMB itself. Bolometers cooled at 0.1 K will allow reaching this sensitivity and, at the same time, reach an angular resolution of 5 minutes of arc. The Low Frequency Instrument (LFI) limited at frequencies less than 100GHz, will use HEMT amplifiers cooled

François R. Bouchet

at 20 Kelvin to increase their sensitivity. The scan strategy is of the total power type. Both instruments use internal references to obtain this total power measurement. This is a 0.1 K heat sink for the bolometers, and a 4 K radiative load for the LFI. The combination of these two instruments on Planck is motivated by the necessity to map the foregrounds in a very broad frequency range: 30 to 850 GHz.

Further details are given in Table I, although these are only indicative since design evolve and in-flight performance might defer from the requirements (in good or bad, either way are possible). Further MAP and PLANCK references (concerning for instance their polarisation sensitivity) may be obtained from the web at:

http://map.gsfc.nasa.gov/m_mm/ob_tech1.html

http://tonno.tesre.bo.cnr.it/Research/PLANCK/Redbook

http://astro.estec.esa.nl/SA-general/Projects/Planck/

which give access (respectively) to the MAP home page, and for PLANCK, the "red book" (the report produced at the end of the phase A study, basis of ESA's selection in 1996), and the Planck Science Team pages. The PLANCK "blue book" should soon appear and it will present "The scientific program of Planck", as of 2002; some of the plots below are extracted from it.

Table I. Summary of (only indicative) experimental characteristics used for comparing experiments. Central band frequencies, ν, are in Gigahertz, the FWHM angular sizes, are in arc minute, and ΔT sensitivities are in μK per $\theta_{FWHM} \times \theta_{FWHM}$ square pixels; the implied noise spectrum normalisation $c_{noise} = \Delta T (\Omega_{FWHM})^{1/2}$, is expressed in $\mu K.deg$.

MAP (as of January 1998)					
ν	22	30	40	60	90
FWHM	55.8	40.8	28.2	21.0	12.6
ΔT	8.4	14.1	17.2	30.0	50.0
c_{noise}	8.8	10.8	9.1	11.8	11.8

Proposed Planck/LFI					Proposed Planck/HFI					
ν	30	44	70	100	100	143	217	353	545	857
FWHM	33	23	14	10	10.7	8.0	5.5	5.0	5.0	5.0
ΔT	4.0	7.0	10.0	12.0	4.6	5.5	11.7	39.3	401	-
c_{noise}	2.5	3.0	2.6	2.3	0.9	0.8	1.2	3.7	38	1711

Figure 9 compares the 1-σ contour on the $C(\ell)$ which we may anticipate for MAP and Planck, *once the foregrounds separation has been performed*, which is an important factor at the level of precision of Planck. That will be further discussed in the foregrounds section.

Another major source of information is the polarisation of the CMB which is entirely generated at the LSS. It is convenient to decompose the polarisation field into two scalar field denoted E and B (to recall the similarity of their parity properties with that of the electromagnetic fields). The power spectrum of the E part is expected to be about 10 times smaller than for the temperature, and the B part (which is only generated by tensor fluctuations) is yet weaker. It is expected that the temperature and E field will be correlated (with a somewhat easier to measure T-E cross power spectrum). The theory for these modes has also been fully developed. Such polarisation measurements may thus be used on their own to constrain the parameters describing the Initial conditions and the cosmological parameters. This will provide (if successful) further checks of the paradigm. Alternatively, polarisation measurements may be used in conjunction with the temperature ones to lift degeneracies between parameters (see for instance Figure 7.b below).

Figure 6 from the "Blue Book" shows the gain of sensitivity to expect between the near-future (Boomerang 2002 & MAP, left) and the Planck (right) experiments for the measurement of the E-type polarisation. One computes similar improvement for the cross-correlation spectrum (between the temperature and E type fluctuations). Planck should also allow a detection of the B spectrum.

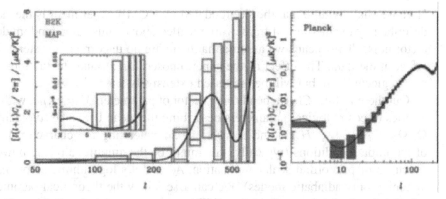

Figure 6. Expected errors on the amplitudes of the E type polarisation for the future Boomerang flight at the end of this year and for MAP on the left, and for Planck on the right.

Such a figure is only illustrative though, since the actual precision reached will depend on how precisely the effect of astrophysical foregrounds fluctuations can be removed. The polarisation signal is expected to be quite weaker than the temperature signal, by at least a factor of ten, and the polarisation properties of the foregrounds are barely known at all. It is therefore quite uneasy if at all possible to assess realistically to what extent foregrounds will decrease our ability at mapping the polarisation of fluctuations at recombination.

Still, the increase in the precision of the determination of the $C(\ell)$'s will be large and one may wonder what might be achieved with that improved sensitivity. Let us suppose the minimal model still (standard inflation) holds; then we shall much improve the precision of the derived parameters (again, if no inconsistency appears). We shall also verify the consistency with other probes of some of the parameters and maybe detect hidden hidden systematic effects in the analyses. More interestingly perhaps, we will have the possibility to consider weak deviations around the minimal model by relaxing priors (prejudice?) on (the absence of) extra degrees of freedom (e.g. isocurvature modes, topological defects, extra scales in the Initial Condition spectrum...) We might also consider more radical deviations as in brane cosmologies (where e.g. $H2 = 8^{\pi}/3(\rho + \rho^2/\sigma)$ for Randall-Sundrum type models), although detailed predictions might turn out difficult to compute. Both approach will be interesting and will certainly be done. To proceed further, I now turn to the approach used for assessing what an experiment can tell on the parameters of a theory.

5. The power of an experiment

Here we shall assume that the sky rendered by a CMB experiment is the sky described by the theoretical framework recalled above, plus some random detector noise. This means we assume that there are no uncorrected systematic effect in the data. The effect of the superimposed foreground fluctuations is also neglected here but will be discussed extensively later.

Our theory of the $C(\ell)$ depends on a vector of parameters $V = (\theta, \mu)$, with θ a series of cosmological parameters describing the mean Universe (typically, Ω, Ω_B, Ω_{CDM}, Ω_Λ, H...) and μ a series parameterising the characteristics of the initial conditions (typically, this might be the amplitude of the power spectrum of primordial scalar fluctuation, A_S, and it's logarithmic slope, n_s, assuming only adiabatic modes). We can assess how the theoretical parameters V (assuming our theory is true) will be constrained by the data by using Bayes theorem which simply states

$$P(V|D)\,P(D) = P(D|V)\,P(V), \tag{2}$$

i.e. in the absence of a theoretical prior within that parameterisation of the theory $(P(V) = 1)$, the probability of the parameters describing the theory

(θ_i, μ_j) given the data is proportional to the probability of the data given the theory (in fact the theory parameters), which is usually called the likelihood.

A classical result is that the Fisher matrix

$$F_{ij} = - < \frac{\partial^2 \mathcal{L}}{\partial T_i \partial T_j} >, \quad \text{with } \mathcal{L} = - \ln \mathcal{P}(\mathcal{D}|\mathcal{T}) \tag{3}$$

gives an indication of the power of the data to constrain the theory. Indeed, the variance of a given theoretical parameter V_i is given by the inverse of the corresponding diagonal element, $\sigma_i^2 = F_{ii}^{-1}$. And even better $F_{ii}^{-1/2}$ gives the variance on the theoretical parameter i after marginalisation over the other parameters (i.e. after integrating over possible values of the other parameters as weighted by their probability distribution function). In other words, the first estimates gives the variances obtained by a joint estimation of all the parameters, while the seconds tell us the constraints on each parameter taken independently.

In the CMB case, the Fisher matrix writes simply

$$F_{ij} = \sum_{\ell} \frac{(2\ell+1)f_{sky}}{2} \{C_\ell + C_N \exp[\theta_B \ell(\ell+1)]\}^{-2} \frac{\partial C\ell}{\partial V_i} \frac{\partial C_\ell}{\partial V_j} \tag{4}$$

where f_{sky} is the fraction of the sky covered by an experiment, so that $(2\ell+1) f_{sky}$ approximates the number of m modes contributing to the analysis at each ℓ. Here C_N fixes the level of the detector noise, assumed white, and θ_B stands for the width of the beam (or angular response) of the experiment, assumed to be Gaussian. This shows that the larger the sky fraction and the lower the noise power spectrum and the beam, the larger will be F_{ij}, and the better will be the constraints. Similarly we find that if the $C(\ell)$ depends weakly on a parameter i (in the exploitable range where $\{C_\ell + C_N \exp[\theta_B \ell(\ell+1)]\}^{-2}$ is sizeable), the corresponding σ_i will be large. Therefore experiments will try to maximise their sky coverage, sensitivity and resolution, with different compromises being made by each team (or different strategies followed) to maximise return at a given time...

Figure 7 shows the expected precision from Planck around two different target models, together with a range of predictions for various inflationary models when their underlying parameters are varied. At left, it is assumed a quite low "true" value of $n_S = 0.9$ and a quite high value for the ratio of the scalar to tensor normalisation factors of the initial power spectra $r = 0.7 = A_S/A_T$. Planck should thus have a rather good discriminatory power in that "easy" case. At right though, it has been assumed a closer value to one for the logarithmic slope of the scalar power spectrum, $n_S = 0.95$, with a much lower tensorial contribution, with $r = 0.005$. In that much less favourable case, Planck should still pin down rather accurately the value of n_S, but will not provide a lower limit to r, leaving alive different models.

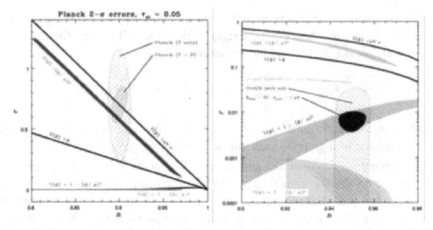

Figure 7. Expected error contours for Planck on the scalar-tensor ratio r and the logarithmic slope n of the Initial fluctuations power spectrum, compared with different models. a) (left) case of a relatively high r and low n b) case (probably more realistic) of a lower r and higher n (note that the inner black contour corresponds to 3 times the nominal sensitivity of Planck). Reprinted from (Kinney, 1998)

It is interesting to note that an increase of sensitivity of 3 would be enough to restrict the possibilities to the black region. This shows that even Planck will not exhaust the information content of the CMB. This plot also shows (light blue contour) what might be expected from a zero noise experiment, but with no polarisation capability, a vivid example of a lift of degeneracy brought by a polarisation measurement.

Before concluding that section, a few cautionary words are in order. Indeed, this Fisher matrix formalism can only tell us the tightest error bars even achievable in an idealistic set up, without even telling us how to reach this optimal. It furthermore assumes that the Likelihood around the true value is Gaussian which is often not such a good approximation.

We should recall too that the power spectra are only a first moment (transform of 2-pt correlation function). While enough to characterise a fully Gaussian distribution, deviations from Gaussianity *are* expected, albeit at a rather low level. Such a detection would reveal much about the mechanisms at work in the early Universe (if they are not residual systematics...).

6. Foregrounds

In the context of the preparation for PLANCK, we developed a model of the statistical characteristics of the various microwave emissions (Bouchet et al., 1997; PLANCK-HFI CONSORTIUM AAO, 1998). One result of this modelling is the angular power spectra, $C(\ell)$, of the fluctuations of all the relevant components, as functions of the frequency. Figure 8.a compares them at

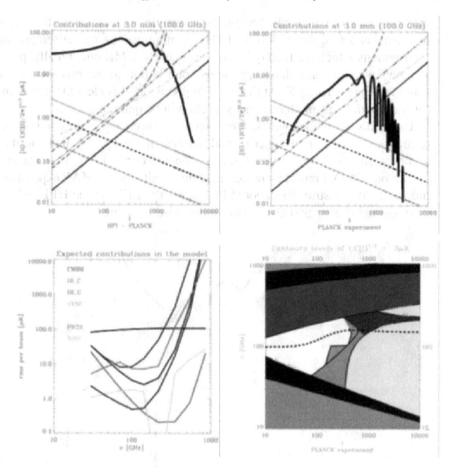

Figure 8. Comparison of sources of microwave anisotropies: a) (Top left) Square root of the contribution to the variance per logarithmic interval, $\ell(\ell+1)C(\ell)$, of the various components at 100 GHz. From top to bottom at $\ell = 10\,000$, one finds the unresolved background from radio-sources, the SZ effect from clusters, the unresolved background from IR galaxies, the CMB (thick line) and the dust, free-free and synchrotron emissions of our galaxy. The 2 other lines that bend upward at large ℓ correspond to the noise spectra of the LFI and the HFI (lower curve). The other panel use the same colour coding. b) (Top right) Foregrounds comparison with the difference between two angular power spectra differing by 2% of the Hubble constant. c) (Bottom left) *rms* contributions of the various components in the PLANCK channels. d) Contours in the frequency-scale plane for the various components.

100 GHz, near the peak of the CMB emission. Panel b) compares these foregrounds power spectra (close to their global frequency minimum) with the difference between two angular power spectra differing by 2% of the Hubble constant. This does show that foregrounds will need be tamed to reach the kind of constraints discussed in the previous section. Panel c) gives the *rms* fluctuation per PLANCK beam which follows by integrating in ℓ the power

spectra of panel a), multiplied by the beam profile, $w_i(\ell)$. It shows that measurements at a frequency close to $100\,\mathrm{GHz}$ minimise the global foreground contributions which are then quite smaller than the CMB one. Finally, panel d) shows when a particular foregrounds contribution to the rms (at a given scale) is greater than $3\mu\mathrm{K}$. At that level of precision, it is clear that removing the contributions from all compact sources will be quite important...

Using this sky model, we can now forecast the accuracy of the components separations for different experimental set-ups, if we specify that the data would be analysed by Wiener filtering (Bouchet et al., 1996; Tegmark and Efstathiou, 1996). Figures 9 and compare the results for the MAP experiment and for the two instruments aboard PLANCK - the LFI and the HFI. All the experimental characteristics used are summarised in table I.

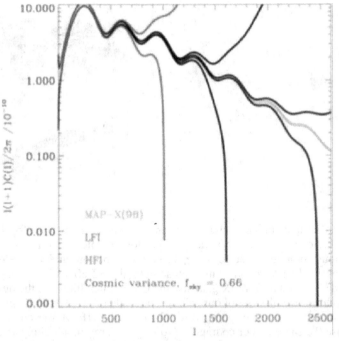

Figure 9. Expected errors on the amplitudes of each mode individually (no band averaging) for different experiments. The thin central lines gives the target theory plus or minus the cosmic variance, for a coverage of 2/3 of the sky. The target theory used here is Lambda CDM (with $\Omega_b = 0.05$, $\Omega_{CDM} = 0.25$, $\Omega_\Lambda = 0.7$, $h = 0.5$). Reprinted from (Bouchet & Gispert, 1999).

If the real sky does not depart too much from the model, and if the experiments deliver the promised performances, Figure 9 indicates the improvements in our knowledge of the CMB spectrum that one may expect in the coming years.

7. Conclusions

The knowledge of CMB anisotropies has literally exploded in the last decade, since their momentous discovery in 1992 by the DMR experiment on the COBE satellite. Since then, the global shape of the spectrum has been uncovered thanks to many ground and balloon experiment, so far confirming the simplest inflationary model and helping shape our surprising view of the Universe: spatially flat, and dominated by dark energy (or Λ) and cold dark matter, with only a few per cent of atoms. But the quest is far from over, with many predictions still awaiting to be checked and many parameters in need of better determination. The next important steps should be, by the end of the year the Archeops results, and early next year those of MAP. In the longer run, balloon and ground experiments will undoubtly start uncovering the properties of the polarised CMB emission. Planck should then definitely fix the temperature power spectrum, provide a very accurate view of the E polarisation, and might measure the B spectrum. Nevertheless a precise determination of the latter will probably have to await a next (fourth) generation satellite dedicated to the measurement of polarisation. In the process, we shall also learn more about astrophysical foregrounds, like clusters, IR galaxies and radio-sources, and the interstellar medium of our own Galaxy.

If the next 10 years are as fruitful as the past decade, many cosmological questions should be settled, from a precise determination of all cosmological parameters to characteristics of the mechanism which seeded the growth of structures in our Universe, if something even more exciting than what is currently foreseen does not surge from the future data...

References

Bouchet, F.R., Gispert, F., Boulanger, R., Puget, J.-P.: 1997, In: *"Microwave Background Anisotropies"*, Proceedings of the 16th Moriond Astrophysics meeting, held in Les Arcs, France, March 16th-23rd, 1996, F.R. Bouchet, R. Gispert, B. Guiderdoni, and J. Trân Thanh Vân (eds.), Editions Frontières, pp. 481.

Bouchet, F.R., Gispert, R.: 1999, *New Astronomy* 4, 443.

Bouchet, F.R., Gispert, R., Puget, J.-L.: 1996, In *Unveiling the Cosmic Infrared Background*; held in 1995 in College Park - MD, Baltimore, Maryland, USA, E. Dwek (ed.), *AIP Conference Proceedings* 348, pp. 255.

Bucher, M., Moodley, K., Turok, N.: 2001, *Phys. Rev. Letters* 87, 191301.

Hanany, S., Ade, P., Balbi, A., Bock, J., Borrill, J., Boscaleri, A., de Bernardis, P., Ferreira, P.G., Hristov, V.V., Jaffe, A.H., Lange, A.E., Lee, A.T., Mauskopf, P.D., Netterfield, C.B., Oh, S., Pascale, E., Rabii, B., Richards, P.L., Smoot, G.F., Stompor, R., Winant, C.D., Wu, J.H.P.: 2000, *ApJL* 545, L5.

Kinney, W.H.: 1998, *Phys. Rev. D* 58, 123506.

PLANCK-HFI CONSORTIUM AAO: 1998, *High Frequency Instrument for the* PLANCK *Mission, a proposal to the European Space Agency*, Proposal submitted in response to the Annoucement of Opportunity of ESA.

Sievers, J.L., Bond, J.R., Cartwright, J.K., Contaldi, C.R., Mason, B.S., Myers, S.T., Padin, S., Pearson, T.J., Pen, U.-L., Pogosyan, D., Prunet, S., Readhead, A.C.S., Shepherd, M.C., Udomprasert, P.S., Bronfman, L., Holzapfel, W.L., May, J.: 2002, *ApJ* submitted; [astro-ph/0205387].
Tegmark, M., Efstathiou, G.: 1996, *MNRAS* **281**, 1297.

Probing Dark Energy with Supernova Searches

Sébastien Fabbro
CENTRA – Centro Multidisciplinar de Astrofísica, Instituto Superior Técnico, Lisbon, Portugal

2003 May 9

Abstract. Four years ago, two teams presented independent analyzes coming from photometry of type Ia supernovae at various distances. The results presented back then shook-up the scientific community: the universe is accelerating with a positive repulsive fluid sometimes called dark energy. Not yet any significant work has disproved the fundamental results, yet some doubt subsists in the assumptions behind the full use of type Ia supernovae as perfect distance indicators. The uncertainty of the evolution problem, the explosion mechanisms and the diversity of the observed light curves properties are often cited problems. All these aspects are now being deeply investigated in to-come or already started supernova searches along with the on-going quest of determining the nature of dark energy. We will present here a brief introduction to the use of type Ia supernova in cosmology, the current status of supernova cosmology as well as an overview of the wide supernova surveys about to begin.

Keywords: supernovae, dark energy, cosmological parameters

1. Relating Energy Densities with Distance Measurements

Measuring distances in the universe is a crucial step to get estimates of cosmological parameters. The greater the distances are, the more they give us access to the geometry and the evolution of the universe. Several ideas to get good cosmological distance indicators have been attempted since 1930. Whereas for nearby objects, there exists a variety of distance methods, for more distant objects, one type of astrophysical object has been successful in converting redshifts to Megaparsecs with some good accuracy: Type Ia Supernovae (SNIa).

Two observations are needed to be able to use SNIa as cosmological distance indicators: the spectrum gives us the redshift z and the identification of candidate, and multi-band photometry to reconstruct the lightcurve $F(t)$ of the supernova. Relating the two observations is done through the *luminosity distance* d_L arising directly from the inverse-squared relation $F(t)=L(t)/4\pi d_L^2(z)$, where $L(t)$ is the intrinsic luminosity. The cosmological parameters are only included in d_L; supernovae observations are therefore "cosmological-model-independent". So far, observations have not shown any other dependence of L but the SNIa epoch t and lightcurve shape parameters, conveniently removed after a standardization process (more on this topic).

With only two fair assumptions of an isotropic and homogeneous universe at very large scales, and Einstein's gravity dictating the energy transfers with

M.J.P.F.G. Monteiro (ed.), The Unsolved Universe: Challenges for the Future, 87-98.

geometrical properties, we can describe how the energy content of the universe influences its single expansion factor $a(t)$. Redshift is interpreted as a cosmological effect through the relation $a(t) = a_0(1+z)^{-1}$. Each component of the universe can be represented by a fluid of density of energy ρ, where the i-index represents non-relativistic matter (m), radiation (r) or dark energy (x). The values of the energy densities rule entirely the time dependence of the expansion factor, by the so-called Friedman equation (see e.g. Peacock, 1999):

$$H^2(a) \equiv \left(\frac{\dot{a}}{a}\right)^2 = \frac{8\pi G}{3}\sum_i \rho_i(a) - \frac{kc^2}{a^2},$$

where we define the Hubble constant H, and $k = \{-1, 0, 1\}$ a constant defining the space curvature of a simple topology (respectively open, flat and closed) Universe. At a critical density $\rho_c = \frac{3H_0^2}{8\pi G}$ where $\sum_i \rho_i = \rho_c$, the universe is flat. For lower values, of the sum, it is open. We further reduce the energy densities as $\Omega_i = \frac{\rho_i}{\rho_c}$. Given these notations, and defining $\Omega_k = \frac{kc^2}{H_0^2}$ with $a_0 = 1$, we re-write our Friedman equation as

$$\sum_i \Omega_i = 1 - \Omega_k.$$

At present, observations show that only the matter and dark energy components are significant, and possibly, as indicated from all latest CMB data (Benett et al., 2003), we could live in a flat universe, leaving simply $\Omega_m + \Omega_x = 1$.

Generally we characterize each energy component of the fluid universe by the equation of state $p_i = w_i \rho_i c^2$, where w_i is the equation of state parameter. We can assume very confidently at many scales that matter behaves as an ideal fluid such that for $w_m = 0$ for non-relativistic matter and $w_r = 1/3$ for relativistic matter and radiation. It becomes less intuitive with the introduction of a repulsive component $w_x < -1/3$, as indicated by observations, very hard to test at laboratory scales. There has been a very extensive literature dealing with the consequences or the nature of a positive, possibly evolving dark energy component. The cosmological constant Ω_Λ is the most simple one. Often referred as a vacuum energy, it could be the relic of some superseding theory. The equation of state parameter keeps a constant value $w_\Lambda = -1$, but recent development introduce a scalar field as a dark energy component, with a $w_x(z)$ evolving with time (see Padmanabhan, 2003, for a review). We will see how it may be possible the reconstruct the equation of state, and we will adopt a first order parameterization $w_x(z) = w_0 + w_1 z$.

Armed with a cosmological description, we can now derive the luminosity distance. It is fairly straight forward to show the following formula (see e.g.

Padmanabhan, 2003):

$$d_L(z) = \frac{c(1+z)}{H_0\sqrt{\Omega_k}} \times$$

$$S\left\{ \sqrt{\Omega_k} \int_0^z dz' \left[\sum_i \Omega_i(1+z')^{3(1+w_i)} - \Omega_k(1+z')^2 \right]^{-\frac{1}{2}} \right\},$$

where $S(x) = \sin(x)$, x or $\sinh(x)$ for closed, flat or open models respectively. Somewhat complicated, the integrand is nothing less than the inverse of the Hubble constant at redshift z' and has an analytical primitive for a flat universe. Figure 1 represents the luminosity distance as a function of redshift , for four different sets of Ω_is. Only for redshifts higher than $z > 0.3$ it is practically possible to discriminate the sets.

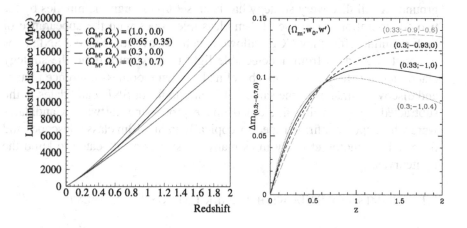

Figure 1. Left panel: Luminosity distance as a function of redshift for various sets of $(\Omega_m, \Omega_\Lambda)$. Only at a redshift of $z \sim 0.3$ and above we can start to discriminate models. *Right panel:* Degeneracies among various cosmological models assuming a flat universe. The combinations of (Ω_m, w_0, w_1) are compared to a $(0.3, -0.7, 0)$ model (Linder & Huterer, 2002). The w' in the figure corresponds to the w_1 in the text.

When we measure supernova fluxes, we have direct access to $H(z)$ and thus a linear combination of the Ω_i, but it causes degeneracies between Ω_i and w_x. Even for pure standard candles, breaking the degeneracy among the full set of involved cosmological parameters $(H_0, \Omega_m, \Omega_x, w_0, w_1)$ requires additional inputs. Typical additional inputs if we are interested in the equation of state parameters, are assuming a flat universe (from CMB), a prior on Ω_m (e.g. from weak lensing), and marginalizing over LH_0. Figure 1 shows the difference in magnitude expected between a flat universe at $\Omega_m = 0.3$, $w_0 = -0.7$ and $w_1 = 0$ with other combinations, but keeping $\Omega_m \sim 0.3$. Nonetheless, to make supernova results as independent as possible, it is very crucial to obtain a large set of well observed and well redshift-sampled supernovae up to a redshift $z > 1.5$. Multiple redshift observations brings more leverage and important degeneracy breaks between the cosmological parameters (Astier, 2001, Goliath et al., 2001, Linder & Huterer, 2002).

2. Current Results

Obtaining a set of supernovae at various redshift is a dedicated task, on which two teams have concentrated their efforts for the last decade. We present their current results here. SNIa only appear few times per millennium per galaxy and can be photo-metered for about 200 days in rest frame at low-z. Hunting supernovae requires both a wide field and large mirror to cover a large volume of space, getting millions of galaxies up to redshift of $z \sim 1$ from ground. A full discovery strategy has been set up through the nineties by the two collaborations. The use of 4-m class telescopes with the appearance of high quantum efficiency CCDs allowed such discoveries. Since a SNIa rises in about ~ 20 days from undetectable light to maximum light, the strategy elaborated was to use a 3 weeks baseline between moonless reference images and discovery images of the same fields, and look for SNIa candidates in the subtracted images, using fine-tuned image processing software. Candidates were whereupon confirmed spectroscopically in 8m-10m class telescopes and followed-up photometrically in as many telescopes as allocated to build the lightcurves.

2.1. DIMMING OF SNIA WITH A DARK ENERGY COMPONENT

Perlmutter et al. (1999) presented cosmological parameters estimations from an analysis of 42 high redshift SNIa, combined with 18 low redshift ones from the Calan-Tololo campaign (Hamuy et al., 1996). The observed apparent luminosity showed a systematic dimming compared to a simple $\Omega_m = 1$, $\Omega_x = 0$ model, which after careful checks, was interpreted as a clear evidence of a presence of dark energy. Riess et al. (1998) showed very similar results from completely independent analysis, from 10 high redshift and 27 low redshift SNIa (within which the Calan-Tololo ones) . If we assume the universe is composed of matter and a cosmological constant, the combined results show the existence of a cosmological constant with 99.9% confidence leading to an accelerating expansion. Figure 2.1 show the combined confidence contours for Ω_m and Ω_Λ for the two teams.

By now assuming a flat universe, and combining measurements, with redshift surveys to constrain Ω_m and CMB measurements, we can get some limits on the equation of state parameter. The constraints only suggest $w_x <$

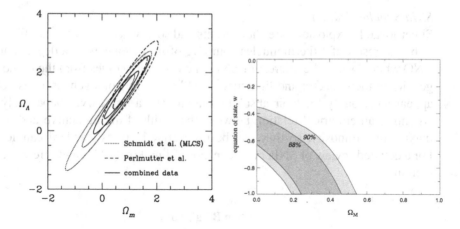

Figure 2. Left panel: Combined confidence contours in the $(\Omega_m, \Omega_\Lambda)$ plane for both high red-shift SNIa results (Wang, 2000). *Right panel:* Equation of state parameter (assumed constant in z) and Ω_m confidence interval, for a flat universe from Perlmutter et al. (1999) results.

-0.5 - see e.g. Perlmutter, Turner & White (1999) - and yet neither per-mit to discriminate among dark energy models, nor permit to reconstruct a dynamical equation of state.

The statistical uncertainties are still the largest contribution to the to-tal error budget, although the systematics are not far behind. With a larger statistical sample, improving the current measurements would definitely re-quire much better control over the systematic uncertainties. Extensive internal checks have been performed on the properties of the high-z and low-z objects used in these measurements in order to detect difference in the samples. We find that systematic errors are mostly due to our limited knowledge of SNIa photometric and spectroscopic behavior. Indeed, when measuring distances using SNIa, we apply corrections for instrumental (e.g. k-corrections) and foreground effects (extinction). The accuracy of these corrections depends heavily on our knowledge of SNIa intrinsic properties. At the moment, a half-dozen significant systematic errors have been identified. We will present the most important ones below.

2.2. PROBLEMS

The dimming of SNIa diminishes down to ~ 0.28 mag at $z \sim 0.5$ compared to an open universe with $\Omega_m = 0.3$, $\Omega_\Lambda = 0$, and several attempts have accounted the dimming for other reasons but the distance. Apart for non-standard cos-mological models, a more direct and physical explanation is to reconsider the assumptions behind the use of SNIa.

SNIa Standardization

Supernovae Ia explosions are short events and somewhat rare. They are likely to be the result of a thermonuclear burning of iron elements of a degenerate CNO white dwarf. The interest of SNIa in cosmology comes from the homogeneity in their spectra and lightcurves. Although the homogeneity does not appear so strongly for their absolute magnitudes, a lightcurve shape analysis shows an empirical relation between the width of the lightcurve and the maximum luminosity, so they can be used effectively as standard candles. For a detailed review of SNIa properties, see Blinnikov (2003) and references therein.

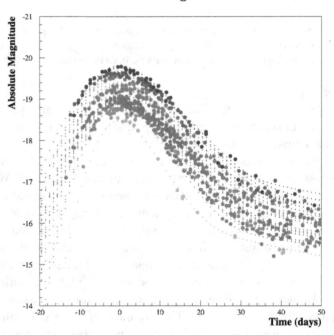

Figure 3. Lightcurve of nearby SNIa in absolute *B* magnitude. We can clearly see the dependence brighter-slower. Brighter lightcurves are actually also a bit bluer (Regnault, 2000).

From accurate photometry, we derive lightcurve properties. With a full selection of observed SNIa, the dispersion at maximum in *B* and *V* reaches $\sigma > 0.25$ mag. But removing a few extraordinary spectra and lightcurves from the set, we note a clear shape resemblance, as shown in Figure 2.2. As it can be seen, the slower supernovae are also the brighter ones (also bluer). This empirical relation has not yet found a satisfactory quantitative answer among radiative transfer models, although most of them suggest the phenomenon depends on the [56]Nimass. The temperature increases with the abundance of ^{56}Ni, accounting for the brighter events, but also increases the opacity. Photons are trapped for a longer time, accounting for the wider lightcurves.

To characterize the brighter-slower relation, several parameterizations are available: the Δm_{15} representing the decrease in magnitude in the B band 15 days following the maximum brightness (Phillips, 1993), the "stretch" of the time-axis of a B-band template light-curve (Goldhaber et al., 2001; Perlmutter et al., 1997), or the Δ parameter of the Multi-Color Light Curve Shape (MLCS), a method using a trained template to correlate simultaneously all colors and absolute magnitudes with a single varying shape (Riess, Press & Kirshner , 1996). The reader is referred in (Leibundgut, 2003) for a critical overview of standardization of the SNIa light curves. Each method has its own application, and lead for all cases to an effective peak magnitude dispersion of $\sigma < 0.2$, which translates into a dispersion on the distance of about 7%.

Extinction

Supernova light can be dimmed by dust present in the optical path. Roughly 10% of them show significant extinction. Usually extinction is color dependent: dimmed objects appear redder and the total absorption is proportional to the reddening. Given supernova intrinsic color, we can correct for reddening. Aguirre (1999) pointed out reddening dust is not the only one and suggested that intergalactic "gray" dust existence is not ruled out. It could be expelled from galaxies, and its possibly large grain size ($> 0.1\mu m$) could produce a very small reddening effect, undetectable on the current observations.

Two possibilities are explored today to measure the importance of gray dust. The first one by taking multi-band photometry to increase the color leverage and make the gray dust more detectable. A first attempt, though not yet conclusive was preformed on one supernova (Riess et al., 2000). The second one is simply going at higher redshift to discriminate between scenarios of gray dust or dark energy. Again, a first attempt at $z \sim 1.7$ based on a single supernova with only photometric measurements (Riess et al., 2001) do not show compatibility with a gray dust universe.

Finally one must not forget that dust correction from our Galaxy is subject to uncertainties and going at higher redshift means going redder and thus less affected by the Milky Way dust. Such a systematic error is today estimated as ~ 0.06 mag.

Evolution

Differences in SNIa do exist. Intrinsic dispersion of lightcurves, colors or spectra show the diversity of the event, although qualitatively similar. It could be due to the environmental effects. In the nearby universe, SNIa in early hosts show narrower lightcurves than late-type hosts (Hamuy, 2000). But after lightcurve shape correction, the dispersion is below observational errors. Similar results were just recently analyzed at high redshift using Hubble

Space Telescope imaging for morphology studies, and Keck spectroscopy (Sullivan et al., 2003). After lightcurve shape and host galaxy corrections, no deviation from a dark energy dominated universe were found, but the Hubble diagram showed more scatter for the SNIa found in late type hosts. We could take an empirical approach and only select sub-sample of SNIa, for example only in elliptical galaxies, to reduce intrinsic dispersion. A more ideal method, is to use a good theoretical modeling of spectra reproducing all SNIa events and instrumental transmission of each band properly calibrated to reproduce lightcurves at any redshift. SNIa spectrum models have unfortunately not yet reached the required precision and practicality for such an evolution-free method to work efficiently with only few parameters.

Two directions have been taken by the observational teams to see on a first order if evolution has its role in the dimming of high redshift supernovae: going at higher redshift to see a departure on Hubble diagram, and increasing the statistical sample with multi-band photometry and spectroscopy.

K-Corrections and others

High redshift SNIa spectra are shifted toward longer wavelengths, and when integrated through the instrumental transmission, K-corrections are needed to recover rest-frame photometry. In order to minimize the error made in extrapolating outside the spectral range, we apply cross-band K-corrections, where we compare the matching nearby SNIa bands with the high redshift ones (Kim, Goobar & Perlmutter, 1996). Nevertheless accuracies on various parameters affect the correction. First, in order to perform the K-correction at any epoch of the supernova, we need a template spectrum, not yet available with required accuracy at all wavelength. Late investigation showed that correlations between stretch, extinction and K-corrections could correct a bit for a lack of template knowledge (Nugent, Kim & Perlmutter, 2002). Other inaccuracies come from calibrations to a standard band-pass system and calibration of supernovae spectra. A good cure to K-corrections will be when we have a large sample of well observed SNIa spectro-photometrically, such as the Nearby Supernova Factory project is about to produce.

Other sources of systematics are shown to be unsignificant , although potentially problematic. Selection biases (such as Malmquist bias) are corrected through monte-carlo studies, but have not been quantified for the used low-redshift sample. Gravitational lensing should affect $z > 1$ supernovae, but has been quantified to be less than 0.02 mag for the published sample at $\bar{z} \sim 0.5$.

More Recent Analyzes

Since the 1998 announcements of the two teams, there has been some progress, observing more supernovae at both nearby and high redshift. One of the achieved development was to get the Hubble diagram against the host galaxy morphology (Sullivan et al., 2003). High resolution images from HST and spectra from Keck were used to a more detailed check on host galaxy effects on the set presented in (Perlmutter et al., 1999). No significant dimming due to extinction was revealed. Also preliminary results from 11 high-redshift SNIa more accurately photo-metered with HST (Knop et al., 2002) as well as other results from (Krisciunas et al., 2003) converge to similar cosmological conclusions as the 1998 results.

3. Prospects

3.1. THE IMPORTANCE OF A NEARBY SUPERNOVA PROGRAM

A better understanding of intrinsic properties of SNIa is of primary importance both in a search for systematic effects and in the precise measurements of cosmological parameters. Such an understanding could arise from precise spectroscopic observations, achievable at low redshift. The *Nearby Supernova Factory* [1] is a project to start in the year to come to detect and follow more than 400 supernovae at a redshift $z < 0.05$. Thanks to a dedicated integral field spectrograph, it would allow to produce data-hyper-cubes (x, y, t, λ) of a supernova and its local surroundings. The wavelength coverage $[0.32, 1]\mu m$ and high resolution $0.3 nm$ of a $6' \times 6''$ field around the target will allow us to directly point some problems addressed in the previous paragraph: comparison with theoretical models, K-corrections, evolution and extinction.

3.2. TOWARD THE NATURE OF DARK ENERGY: ONE STEP FURTHER

Going further into precision supernova cosmology will not only require better knowledge of the supernova event, but also a larger homogeneous statistical sample of high redshift supernovae. One of the main problems associated with high redshift supernova campaigns is managing, reduce and inter-calibrate all observations from the various telescopes used for follow-up photometry. Apart from regular searches organized every semester by the SCP and High-Z Team, two completely dedicated project aim to observe few hundreds supernovae at redshift between $0.1 < z < 0.9$: the SuperNova Legacy Survey (SNLS) [2] and the Essence project [3].

The SNLS is one of the CFHT Legacy Survey that will start February 2003. It uses the Megaprime imager, a wide-field camera (Megacam) mounted on the prime focus of the CFHT-3.6m telescope in Hawaii. The camera is a mosaic of 36 thinned 2K×4.5K CCDs covering one square degree. Four fields will be continuously observed for the next 5 years, in four bands (g, r, i, z).

---

[1] see http://snfactory.in2p3.fr

[2] see http://snls.in2p3.fr

[3] see http://www.ctio.noao.edu/wproject

The observation strategy has thus been adapted, now allowing early discovery and well-sampled homogeneous follow-up photometry of 600 SNIa in 5 years. Such a tremendous data set offer the possibility of checking the standardization technique at various redshifts, and on different filters.

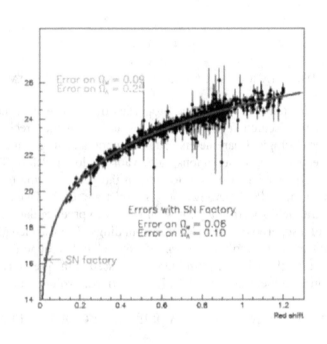

Figure 4. Expected Hubble diagram after 5 years of SNLS survey.

To reach better than 10% precision on Ω_m and Ω_x, it will be necessary to combine the SNLS measurements with the Nearby Supernova Factory ones. Together with a prior on Ω_m from the CMB and the CFHT weak lensing survey, should lead to an uncertainty on w_x of $\sigma \sim 0.1$.

3.3. UNAMBIGUOUSLY UNVEILING THE DARK ENERGY

Going at higher redshift and beating 2% measurements of cosmological parameters will require a very large survey for well measured high redshift SNIa. Only space observations can give us good photometric precision and spectroscopy at high redshift, where most of the SNIa flux is in infrared. The *SuperNova Acceleration Probe* (SNAP) [4] is a project of a 2-m telescope in space, mounted with a wide field imager, an infrared imager and an integral

---

[4] see http://snap.lbl.gov

field spectrometer. The observation strategy is similar to the SNLS, but covering 15 square degrees and reaching 2000 well followed SNIa/year, with an expected photometric precision of less than 2%.

SNAP data quality should be able to discriminate among dark energy models and alternative explanations to the acceleration of universe expansion. In principle, it would be able to detect time variation of the equation of state, together with prior information on Ω_m and low-redshift SNIa from the Nearby Supernova Factory.

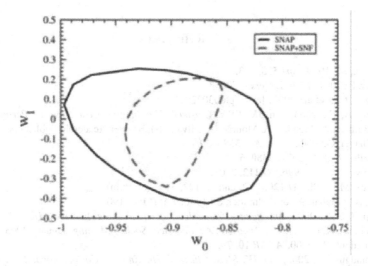

Figure 5. Expected confidence contours (68%) in the (w_0, w_1) plane for the SNAP experiment when the Nearby Supernova Factory SNIa are added (in red-dashed) and when they are not. A flat universe has been assumed with a Gaussian prior on Ω_m of $\sigma = 0.04$.

The SNAP design is still evolving. Depending on the availability of the funds, SNAP is expected to be launched in 2010 and its mission should last at least for three years. It is also designed to do a whole range of science, from galaxy structure to weak lensing.

4. Conclusion

Dark energy as a solution of dimming of type Ia supernovae at high redshift is almost a secure fact, but as always, it needs further studies. Current constraints on the nature of dark energy from SNIa can be greatly improved with a much better control of systematics, requiring precise nearby supernova observations, and by increasing the sample of large high redshift supernovae. Few projects directly aim at building such samples. If type Ia supernovae keep holding their wonderful standardization properties, we can expect within five years to have a first measurement the dark energy equation of state with a precision of about 10%.

Acknowledgments

S. Fabbro thanks the support for this work provided by the Fundação para a Ciência e Tecnologia through project PESO/P/PRO/15139/99 and a Fellowship grant through Project ESO/FNU/43749/2001.

References

Aguirre, A.: 1999, *ApJ* **525**, 583.
Astier, P.: 2001, *Phys. Letters B* **500**, 8.
Bennett, C.L. et al.: 2003, [astro-ph/0302207].
Blinnikov, S.: 2003, in *3K, SN's, Clusters: Hunting the Cosmologocal Parameters*, D. Barbosa, L. Mendes, A. Mourão, A. Silva (eds), Kluwer Academic Publishers.
Goldhaber, G. et al.: 2001, *ApJ* **558**, 359.
Goliath, M. et al.: 2001, å**380**, 6.
Hamuy, M. et al.; 1996, *AJ* **112**, 2408.
Hamuy, M.: 2000, *AJ* **120**, 1479 and *AJ* **122**, 3106 (erratum).
Kim, A., Goobar, A., & Perlmutter, S.: 1996, *PASP* **108**, 190.
Knop, R. et al.: 2002, *American Astronomical Society Meeting*, January 2002.
Krisciunas, K. et al.: 2003, *American Astronomical Society Meeting*, January 2003.
Leibundgut, B.: 2000, *A&AR* **10**, 79.
Leibundgut, B.: 2003, in *3K, SN's, Clusters: Hunting the Cosmologocal Parameters*, D. Barbosa, L. Mendes, A. Mourão, A. Silva (eds), Kluwer Academic Publishers.
Linder, E., & Huterer, D.: 2002, *Phys. Rev. D* **67**, 081303.
Nugent, P., Kim, A., & Perlmutter, S.: 2002, *PASP* **114**, 803.
Padmanabhan, T.: 2002, [hep-th/0212290].
Peacock, J.: 1999, in *Cosmological physics*, Cambridge University Press, 1999.
Perlmutter, S. et al.: 1997, *ApJ* **483**, 565.
Perlmutter, S. et al.: 1999, *ApJ* **517**, 565.
Perlmutter, S., Turner, M., White, M.: 1999, *Phys. Rev. Letters* **83**, 670.
Phillips, M.: 1993, *ApJL* **413**, 105.
Regnault, N.: 2000, ijn *Recherche de Supernovae avec EROS2. Etude Photométrique de SNIa Proches et Mesure de H_0*. PhD Thesis, University Denis Diderot, Paris VII.
Riess, A., Press, W., & Kirshner, R.: 1996, *ApJ* **473**, 88.
Riess, A. et al.: 1998, *AJ* **116**, 1009.
Riess, A. et al.: 1999, *AJ* **117**, 707.
Riess, A. et al.: 2000, *ApJ* **536**, 62.
Riess, A. et al.: 2001, *ApJ* **560**, 49.
Sullivan, M. et al.: 2003, *MNRAS* **340**, 1057.
Wang, Y.: 2000, *ApJ* **546**, 531.

Cosmological Constraints from Chandra X-ray Observations of Galaxy Clusters

Steven W. Allen

Institute of Astronomy, University of Cambridge, UK

2002 November 25

Abstract. CHANDRA X-ray observations of rich, dynamically relaxed galaxy clusters allow the properties of the X-ray gas and the total gravitating mass to be determined precisely. Here, we discuss how CHANDRA observations may be used as a powerful tool for cosmological studies. By combining CHANDRA X-ray results on the X-ray gas mass fractions in clusters with independent measurements of the Hubble constant and the mean baryonic matter density of the universe, we obtain a tight constraint on the mean total matter density of the universe, Ω_m, and an interesting constraint on the cosmological constant, Ω_Λ. Using these results, together with the observed local X-ray luminosity function of the most X-ray luminous galaxy clusters, a mass-luminosity relation determined from CHANDRA and ROSAT X-ray data and weak gravitational lensing observations, and the mass function predicted by numerical simulations, we obtain a precise constraint on the normalization of the power spectrum of density fluctuations in the nearby universe, σ_8. We compare our results with those obtained from other, independent methods.

Keywords: cosmological parameters – X-rays: galaxies: clusters – gravitational lensing — large-scale structure of the universe — X-rays: galaxies: clusters

1. Introduction

Accurate measurements of the masses of clusters of galaxies are of profound importance to cosmological studies. Galaxy clusters are the largest gravitationally-bound objects known and their spatial distribution, redshift evolution and matter content provide sensitive probes of cosmology.

Currently, our two best methods for measuring the masses of clusters use X-ray observations and studies of gravitational lensing by clusters. X-ray mass measurements are based on the assumption that the hot ($\sim 10^8$ K) gas in clusters is in hydrostatic equilibrium. The distribution of mass can be determined once the temperature and density profiles of the X-ray gas are known. The launch of the CHANDRA X-ray Observatory and XMM-NEWTON have revolutionized X-ray studies, permitting the first detailed, spatially-resolved X-ray spectroscopy of clusters and opening the door to precise mass measurements for dynamically relaxed systems from X-ray data. Gravitational lensing studies, in contrast, offer a method for measuring the masses of clusters that is essentially free from assumptions about the dynamical state of the gravitating matter, allowing these techniques to be applied to dynamically active, merging systems. The last decade has seen rapid progress in lensing studies,

M.J.P.F.G. Monteiro (ed.), The Unsolved Universe: Challenges for the Future, 99-108.

Steven W. Allen

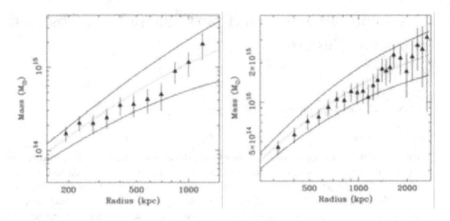

Figure 1. A comparison of the mass measurements obtained from CHANDRA X-ray observations (solid lines) and wide field weak lensing studies (triangles) of two of the dynamically relaxed clusters in our sample: Abell 2390 (left; Allen et al., 2001) and RXJ1347.5-1145 (right; Allen et al., 2002b). Error bars are 68 per cent confidence limits.

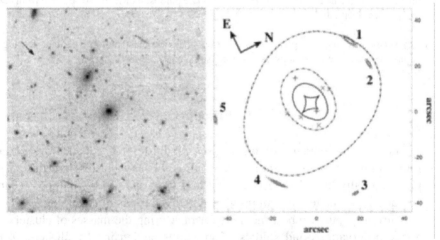

Figure 2. The observed (left) and predicted (right) gravitational arc geometry in RXJ1347.5-1145 (Allen et al., 2002b). The source positions in the model are denoted by crosses. The central positions for the two dominant mass clumps in the cluster are marked with plus signs. Also shown are the critical curves (dash-dotted) and caustic lines for a source at a redshift $z = 1.0$.

to the point where obtaining weak lensing mass measurements for the largest clusters has become a relatively straightforward task (e.g. Dahle et al., 2002). The combination of X-ray and gravitational lensing observations provides the best way to obtain precise, robust measurements of the masses of clusters, as well as a powerful probe of the dynamical states of these systems.

2. X-ray and lensing mass measurements

The methods employed in the X-ray and gravitational lensing analyses are described by Allen et al. (2001; 2002b) and Schmidt et al. (2001). The CHANDRA observations were made using the Advanced CCD Imaging Spectrometer (ACIS). The weak lensing results are drawn from the literature and use ground-based optical observations. The strong lensing analyses have been carried out using our own codes and (primarily) data from the Hubble Space Telescope. For details of the sample see Allen et al. (2003).

Figure 1 shows a comparison of the CHANDRA X-ray and weak lensing results on the total mass profiles (luminous plus dark matter) for two of the dynamically relaxed clusters in our sample: Abell 2390 (left) and RXJ1347.5-1145 (right) – the most X-ray luminous cluster known. We find that for both clusters the mass profiles can be parameterized using the 'universal' form determined from simulations by Navarro, Frenk & White (1997), with concentration parameters and scale radii within the ranges expected for clusters of these masses. The agreement between the independent lensing and X-ray mass measurements is good in both cases, confirming the validity of the hydrostatic assumption used in the X-ray analysis and ruling out significant non-thermal pressure support on these spatial scales.

Figure 2 shows the observed (left) and predicted (right) gravitational arc geometry in RXJ1347.5-1145 (Allen et al., 2002b). We find that a simple two-component mass model, with ellipticities and orientations for the mass components matching those of the dominant cluster galaxies, provides a reasonable description for the overall arc geometry in the cluster. Such a model is consistent with the Chandra X-ray and weak lensing results. Similar results are obtained for other clusters e.g.Abell 1835 (Schmidt et al., 2001) and MS2137.3-2353 (Schmidt & Allen, 2003, in preparation).

3. Cosmological constraints from the X-ray gas mass fraction

The matter content of rich clusters of galaxies is thought to provide an almost fair sample of the matter content of the universe as a whole (White et al., 1993). The observed ratio of the baryonic to total mass in clusters should therefore closely match the ratio of the cosmological parameters Ω_b/Ω_m, where Ω_b and Ω_m are the mean baryon and total mass densities of the Universe in units of the critical density. The combination of robust measurements of the baryonic mass fraction in clusters with accurate determinations of Ω_b from cosmic nucleosynthesis calculations (constrained by the observed abundances of light elements at high redshifts) can therefore be used to determine Ω_m.

This method for measuring Ω_m, which is particularly simple in terms of its underlying assumptions, was first highlighted by White & Frenk (1991) and subsequently employed by many groups (e.g. David, Jones & Forman, 1995; White & Fabian, 1995; Evrard, 1997; Ettori & Fabian, 1999). The baryonic mass content of rich clusters of galaxies is dominated by the X-ray emitting intracluster gas, the mass of which exceeds the mass of optically luminous material by a factor ~ 6; the baryonic mass fraction in stars, determined from detailed studies of nearby clusters, scales as $0.19h^{0.5}f_{gas}$, where f_{gas} is the fraction of the total mass in X-ray gas (e.g.White et al., 1993; Fukugita, Hogan & Peebles, 1998.) Since the X-ray emissivity of the X-ray gas is proportional to the square of its density, the gas mass profile can be precisely determined from the X-ray data. With the advent of accurate measurements of Ω_b (e.g. O'Meara et al., 2001 and references therein) and a precise determination of the Hubble Constant, H_0 (Freedman et al., 2001), the dominant uncertainty in determining Ω_m from the baryonic mass fraction in clusters had lain in the measurements of the total (luminous plus dark) matter distributions in the clusters. However, the new CHANDRA data and gravitational lensing studies have now reduced the uncertainties in the baryonic mass fraction measurements to $\lesssim 10$ per cent, an accuracy comparable to the current Ω_b and H_0 results.

Expanding on the simple arguments outlined above, we can obtain more rigorous constraints on cosmological parameters if we also examine the apparent variation of the observed f_{gas} values with redshift. Such methods were first proposed by Sasaki (1996) and Pen (1997). The key point is that when measuring the X-ray gas mass fraction from the X-ray data, one needs to adopt a reference cosmology. The measured f_{gas} value for each cluster will depend upon the assumed angular diameter distance to the source as $f_{gas} \propto D_A^{1.5}$. Thus, although we expect the observed f_{gas} values to be invariant with redshift, this will only *appear* to be the case when the assumed cosmology matches the true, underlying cosmology.

Figure 3 shows the measured f_{gas} values as a function of redshift for an assumed Einstein-de-Sitter (SCDM: $\Omega_m = 1.0$, $\Omega_\Lambda = 0.0$) cosmology, with a Hubble parameter $h = H_0/100 \text{km s}^{-1} \text{Mpc}^{-1} = 0.5$. The results indicate an apparent drop in f_{gas} as the redshift increases. In order to obtain our constraints on cosmological parameters, we have fitted the data in Figure 3 with a model which accounts for the expected apparent variation in the $f_{gas}(z)$ values as the true, underlying cosmology is varied. In detail, the fitted model function is

$$f_{gas}^{mod}(z) = \frac{b\Omega_b}{\left(1 + 0.19\sqrt{h}\right)\Omega_m}\left[\frac{h}{0.5}\frac{D_A^{\Omega_m=1,\Omega_\Lambda=0}(z)}{D_A^{\Omega_m,\Omega_\Lambda=1-\Omega_m}(z)}\right]^{1.5} \tag{1}$$

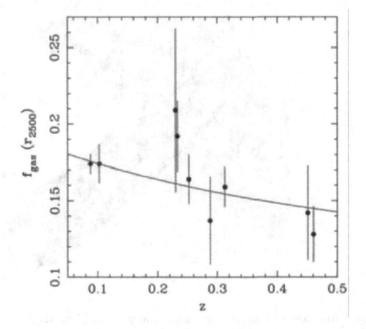

Figure 3. The apparent redshift variation of the X-ray gas mass fraction measured within a density contrast $\Delta = 2500$ (with rms 1σ errors) for the sample of dynamically relaxed clusters observed with CHANDRA. (An SCDM reference cosmology with $h = 0.5$ is assumed; see text.) The curve shows the predicted $f_{gas}(z)$ behaviour for the best-fitting cosmology with $\Omega_m = 0.292$ and $\Omega_\Lambda = 0.68$.

which depends on Ω_m, Ω_Λ, Ω_b, h and a bias factor b. The ratio $(h/0.5)^{1.5}$ accounts for the change in the Hubble constant between the considered model and default SCDM cosmology ($h = 0.5$), and the ratio of the angular diameter distances accounts for deviations in the geometry of the Universe from the Einstein-de-Sitter case. The bias parameter b is motivated by gasdynamical simulations which suggest that the baryon fraction in clusters may be depressed slightly with respect to the universe as a whole (e.g. Eke, Navarro & Frenk, 1998; Bialek et al., 2001). We include a Gaussian prior on the bias factor, $b = 0.93 \pm 0.05$, a value appropriate for hot ($kT > 5$ keV), massive clusters in the redshift range $0 < z < 0.5$ from the simulations of Bialek et al. (2001). We also include Gaussian priors on the Hubble constant, $h = 0.72 \pm 0.08$, the final result from the Hubble Key Project reported by Freedman et al. (2001), and $\Omega_b h^2 = 0.0205 \pm 0.0018$ (O'Meara et al., 2001), from cosmic nucleosynthesis calculations constrained by the observed abundances of light elements at high redshifts.

We have examined a grid of cosmologies covering the plane $0.0 < \Omega_m < 1.0$ and $0.0 < \Omega_\Lambda < 1.5$. The joint 1, 2 and 3 σ confidence contours on Ω_m and Ω_Λ are shown in Figure 4. The best-fit parameters and marginalized 1σ

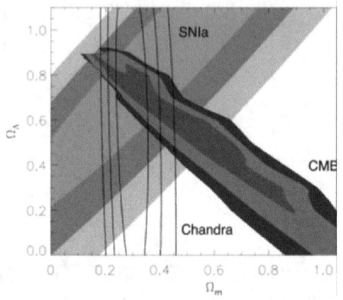

Figure 4. The joint 1, 2 and 3 σ confidence contours on Ω_m and Ω_Λ determined from the CHANDRA $f_{gas}(z)$ data (bold contours), and independent analyses of cosmic microwave background (CMB) anisotropies and the properties of distant supernovae.

error bars are $\Omega_m = 0.292^{+0.040}_{-0.036}$ and $\Omega_\Lambda = 0.68^{+0.42}_{-0.52}$, with $\chi^2_{min} = 4.2$ for 7 degrees of freedom, indicating that the model provides a good description of the data.

Also shown in Figure 4 are the constraints on Ω_m and Ω_Λ obtained by Jaffe et al.(2001) from studies of cosmic microwave background (CMB) anisotropies (incorporating the COBE Differential Microwave Radiometer, BOOMERANG-98 and MAXIMA-1 data of Bennett et al., 1996; de Bernardis et al., 2000; and Hanany et al., 2000, respectively) and the properties of distant supernovae (incorporating the data of Riess et al., 1998; and Perlmutter et al., 1999). The agreement between the results obtained from the independent methods is striking: all three data sets are consistent at the 1σ confidence level, and favour a model with $\Omega_m = 0.3$ and $\Omega_\Lambda = 0.7 - 0.8$. These results are also consistent with the findings of Efstathiou et al. (2002), Lahav et al. (2002), Percival et al. (2002) and Melchiorri & Silk (2002), from combined analyses of more recent CMB and 2dF Galaxy Redshift Survey data. For more details of this work see Allen et al. (2002a,c).

4. The normalization of mass fluctuations in the Universe

By combining CHANDRA results on the masses of dynamically relaxed clusters and wide field weak lensing studies of more dynamically active systems

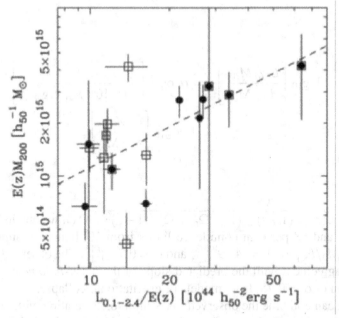

Figure 5. The observed mass-luminosity relation measured within a density contrast $\Delta = 200$ (Allen et al., 2003). CHANDRA mass measurements for dynamically relaxed clusters are indicated by filled circles. Weak lensing mass measurements are indicated by open squares. The four clusters with CHANDRA mass measurements and consistent weak lensing results are indicated by filled circles surrounded by open squares. The best-fitting power law model is shown as the solid curve. The two most significant outliers above (Abell 1351) and below (Abell 209) the best-fit curve appear to be undergoing major merger events. A flat ΛCDM cosmology with $\Omega_m = 0.3$ and $\Omega_\Lambda = 0.7$ is assumed.

with ROSAT measurements of the X-ray luminosities of clusters, we can create a mass-luminosity relation. This provides another powerful tool for cosmological work.

The observed mass-luminosity relation for the 17 X-ray luminous clusters in our current sample, measured within radii r_{200} (corresponding to a density contrast $\Delta = 200$ with respect to the critical density of the universe at the redshifts of the clusters), is shown in Figure 5. Dynamically relaxed clusters with mass measurements from CHANDRA X-ray data are indicated by filled circles. The more dynamically active systems with mass measurements from weak lensing are marked with open squares. The four clusters with CHANDRA mass measurements and consistent weak lensing results are indicated by filled circles surrounded by open squares.

Fitting these data with a power law model of the form

$$\log_{10}\left[\frac{E(z)\,M_{200}}{h_{50}^{-1}\,\mathrm{M}_\odot}\right] = \alpha\log_{10}\left[\frac{L}{E(z)\,10^{44}\,h_{50}^{-2}\,\mathrm{erg\,s^{-1}}}\right]$$

$$+ \log_{10}\left[\frac{M_0}{h_{50}^{-1}\,\mathrm{M}_\odot}\right], \tag{2}$$

where $E(z) = (1+z)\sqrt{(1+z\Omega_m + \Omega_\Lambda/(1+z)^2 - \Omega_\Lambda)}$, we obtain best-fitting values and 68 per cent confidence limits from 10^6 bootstrap simulations of $\log_{10}(M_0/h_{50}^{-1}\,M_\odot) = 14.29^{+0.20}_{-0.23}$ and $\alpha = 0.76^{+0.16}_{-0.13}$. The observed slope is in good agreement with the predicted slope of the mass–*bolometric* luminosity relation of $\alpha = 0.75$, from models of gravitational collapse.

We can combine the observed mass-luminosity relation with the predicted mass function of clusters from the Hubble Volume simulations of Evrard et al. (2002) to predict the X-ray luminosity function (XLF) of clusters (the number density as a function of luminosity) as a function of cosmological parameters: σ_8 – the root-mean-square variation of the linearly-evolved density field smoothed by a top hat window function of size $8h^{-1}$Mpc – and Ω_m. These predictions can then be compared with the observed cluster XLF, determined from the ROSAT All-Sky Survey (Ebeling et al., 2000; Böhringer et al., 2002), to constrain the cosmological parameters.

Figure 6. shows the constraints on σ_8 and Ω_m (for an assumed flat ΛCDM cosmology) obtained by this method, also folding in the constraints on Ω_m from the X-ray gas mass fraction data discussed in Section 3. We obtain the best-fit results $\sigma_8 = 0.695 \pm 0.042$ and $\Omega_m = 0.287 \pm 0.036$ (marginalized 68 per cent confidence limits).

Figure 6 also shows how our results compare with current constraints from analyses of cosmic microwave background (CMB) anisotropies and the 2dF galaxy redshift survey data. We see that the results from all three data sets are consistent at the 68 per cent confidence level. Our results are also in good agreement with the independent analyses of CMB+2dF data reported by Lahav et al. (2002), Percival et al. (2002) and Melchiorri & Silk (2002), the results from measurements of evolution in the cluster XLF reported by Borgani et al. (2001), and some recent findings from studies of cosmic shear (e.g. Jarvis et al., 2002). For more details of this work see Allen et al. (2003).

The constraints on cosmological parameters and, in particular, the constraints on dark energy, should improve rapidly as further CHANDRA, XMM-NEWTON and high-quality gravitational lensing data are gathered.

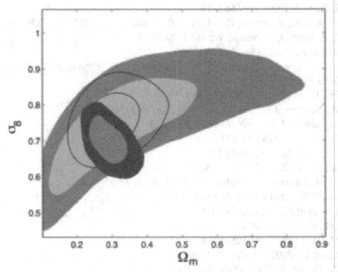

Figure 6. The joint 1 and 2 sigma confidence contours on σ_8 and Ω_m from the present study of X-ray and lensing cluster data (inner, shaded contours: the analysis includes Gaussian priors of $h = 0.72 \pm 0.08$, $b = 0.93 \pm 0.05$ and $\Omega_b h^2 = 0.0205 \pm 0.0018$) together with the results obtained from analyses of current CMB and 2dF Galaxy Redshift Survey data. For the CMB and 2dF analyses, we have used the published Markov Chain Monte Carlo samples of Lewis & Bridle (2002). The CMB data consist of a combination of COBE (Bennett et al., 1996), Boomerang (Netterfield et al., 2002), Maxima (Hanany et al., 2000), DASI (Halverson et al., 2002), Cosmic Background Imager (Pearson et al., 2002) and Very Small Array (Scott et al., 2003) data. The 2dF galaxy redshift survey constraints are from Percival et al. (2002). The results obtained from the CMB data alone, using the 6 parameter model of Lewis & Bridle (2002) and fixing the optical depth to reionization $\tau = 0.04$ and marginalizing over h, $\Omega_b h^2$ and n, are shown as the outer, filled contours. The results obtained from the combined CMB+2dF data set, marginalizing over the same parameters, are shown as solid lines. A flat ΛCDM cosmology is assumed.

Acknowledgements

I thank my collaborators Robert Schmidt, Andy Fabian and Harald Ebeling and the conference organizers for a highly enjoyable meeting.

References

Allen, S.W., Ettori, S., Fabian, A.C.: 2001, *MNRAS* **324**, 877
Allen, S.W., Schmidt, R.W., Fabian, A.C.: 2002a, *MNRAS* **334**, L11
Allen, S.W., Schmidt, R.W., Fabian, A.C.: 2002b, *MNRAS* **335**, 256
Allen, S.W., Schmidt, R.W., Fabian, A.C., Ebeling H.: 2003, *MNRAS* **342**, 287
Bennett, C. et al.: 1996, *ApJ* **464**, L1
Bialek, J.J., Evrard, A.E., Mohr, J.J.: 2002, *ApJ* **578**, L9
Böhringer, H., et al.: 2002, *ApJ* **566**, 93

Borgani, S., et al.: 2001, *ApJ* **561**, 13

Dahle, H., Kaiser, N., Irgens, R.J., Lilje, P.B., Maddox, S.: 2002, *ApJS* **139**, 313

David, L.P., Jones, C., Forman, W.: 1995, *ApJ* **445**, 578

de Bernardis, P., et al.: 2000, *Nat* **404**, 955

Ebeling, H., Edge, A.C., Allen, S.W., Crawford, C.S., Fabian, A.C., Huchra, J.P.: 2000, *MNRAS* **318**, 333

Efstathiou, G., et al.: 2002, *MNRAS* **330**, 29

Eke, V.R., Navarro, J.F., Frenk, C.S.: 1998, *ApJ* **503**, 569

Ettori, S., Fabian, A.C.: 1999, *MNRAS* **305**, 834

Evrard, A.E.: 1997, *MNRAS* **292**, 289

Evrard, A.E., et al.: 2002, *ApJ* **573**, 7

Freedman, W., et al.: 2001, *ApJ* **553**, 47

Fukugita, M., Hogan, C.J., Peebles, P.J.E.: 1998, *ApJ* **503**, 518

Halverson, N.W., et al.: 2002, *ApJ* **568**, 38

Hanany, S., et al.: 2000, *ApJ* **545**, L5

Jaffe, A.H., et al.: 2001, Phys. Rev. Lett. **86**, 3475

Jarvis, M., et al.: 2002, *AJ* **125**, 1014

Lahav, O., et al.: 2002, *MNRAS* **333**, 961

Lewis, A., Bridle, S.: 2002, *Phys. Rev. D* **66**, 103511

Melchiorri, A., Silk, J.: 2002, *Phys. Rev. D* **66**, 041301

Navarro, J.F., Frenk, C.S., White, S.D.M.: 1997, *ApJ* **490**, 493

Netterfield, C.B. et al.: 2002, *ApJ* **571**, 604

O'Meara, J.M., et al.: 2001, *ApJ* **552**, 718

Pearson, T.J., et al.: 2002, *ApJ* submitted [astro-ph/0205388]

Pen, U.: 1997, New Ast. **2**, 309

Percival, W.J., et al.: 2002, *MNRAS* **337**, 1068

Perlmutter, S., et al.: 1999, *ApJ* **517**, 565

Riess, A.G., et al.: 1998, *AJ* **116**, 1009

Sasaki, S.: 1996, *PASJ* **48**, L119

Schmidt R.W., Allen S.W., Fabian A.C.: 2001, *MNRAS* **327**, 1057

Scott, P.F., et al.: 2003, *MNRAS* **341**, 1076

White, D.A., Fabian, A.C.: 1995, *MNRAS* **273**, 72

White, S.D.M., Frenk, C.S.: 1991, *ApJ* **379**, 52

White, S.D.M., Navarro, J.F., Evrard, A.E., Frenk, C.S.: 1993, *Nat* **366**, 429

Cosmological Parameter Estimation with the Galaxy Cluster Abundance

Pedro T. P. Viana
Centro de Astrofísica da Universidade do Porto
Departamento de Matemática Aplicada da Faculdade de Ciências da U.P., Portugal

2003 June 17

Abstract. Clusters of galaxies are the most massive virialized structures in the Universe. Given that the mass function of large-scale structures decreases exponentially at the high-mass end, galaxy clusters are a sensitive probe of its normalization and redshift evolution, and hence of the cosmological parameters that most influence it. It will be discussed to what extent these cosmological parameters, namely the present amplitude of density perturbations, the matter density and a possible cosmological constant, can be constrained using observational data on the present and past abundance of galaxy clusters. Results will be presented based on the available data, as well as expected constraints from the *X-ray Cluster Survey* (**XCS**).

Keywords: cosmology, clusters of galaxies

1. Introduction

Clusters of galaxies are one of the most important probes of the large scale structure and overall dynamical state of the Universe. Their present-day average statistical properties, and their evolution as a function of redshift, can be used to constrain the cosmological parameters that most influence the formation and evolution of large scale structures: the normalization of the power spectrum of density fluctuations, usually given as σ_8 - the dispersion of the density field on scales of $8\,h^{-1}$ Mpc (h is the present value of the Hubble parameter, H_0, in units of $100\ \mathrm{km\,s^{-1}\,Mpc^{-1}}$); the total matter density in units of the critical density, Ω_0; the energy density associated with a possible cosmological constant, Ω_Λ. Recently, this last quantity as been often substituted by Ω_w, where w is a constant assumed to describe the behaviour with time of the equation of state, $w = p/\rho$, of a possible dark energy component ($w = -1$ in the case of a classical cosmological constant, Λ).

We will concentrate on the cluster number density: its present-day value and evolution with redshift. We start by briefly describing why and how it can be used to constrain cosmological parameters. Next, we use recently published data on the local cluster abundance to determine σ_8, and review the estimation of Ω_0 based on the redshift evolution of the cluster abundance. Finally, we discuss future prospects in this field, and determine to what extent the recently begun *XCS* will be able to constrain cosmological parameters.

M.J.P.F.G. Monteiro (ed.), The Unsolved Universe: Challenges for the Future, 109-118.

Pedro T. P. Viana

2. Cosmology with the cluster abundance

The number density of dark matter haloes as a function of mass M and red-shift z, also known as the halo mass function, has long been considered an essential prediction of any credible large-scale structure formation model. Until recently, the Press-Schechter approximation (Press & Schechter, 1974) was widely used to estimate the halo mass function on scales that are still evolving (quasi-)linearly, first only in the context of structure formation models with primordial Gaussian distributed matter density perturbations, but recently extended also to the non-Gaussian case (e.g. Inoue & Nagashima, 2002; and references therein). Predicting the halo mass function on scales that are already evolving non-linearly is much more difficult, and currently one still needs to resort to fully N-body simulations to determine the halo mass function with accuracy. In what follows we will always assume that we are working in the (quasi-)linear regime.

The (comoving) halo mass function can be written as

$$n(M, z) = -F(\sigma)\frac{\rho_0}{M\sigma}\frac{d\sigma}{dM}, \tag{1}$$

with $M(R) = (4/3)\pi R^3\rho_0$, where $\rho_0 = \Omega_0\rho_c$ is the present-day total matter density (ρ_c is the critical density), and

$$\sigma(R,z) = \sigma(R,0) * \frac{g[\Omega_m(z), \Omega_w(z), w]}{g[\Omega_0, \Omega_w(0), w]} * (1+z)^{-1}, \tag{2}$$

is the dispersion of the density field at some scale R at redshift z [g is a function that describes the growth of density perturbations in the linear regime, i.e. as long as $\sigma(R,z) \lesssim 1$]). The density parameters $\Omega_m(z)$ and $\Omega_w(z)$ depend only on Ω_0, $\Omega_w(0)$ and w.

Often, in the case of Cold Dark Matter (CDM) dominated structure formation models, the following parametrization is used

$$\sigma(R,0) = \sigma_8 \left(\frac{R}{8h^{-1}\,Mpc}\right)^{-f(R,\sigma,\Gamma)}, \tag{3}$$

where Γ is a shape parameter, and f is numerically calculated (Viana & Liddle, 1996; Viana & Liddle, 1999).

In the Press-Schechter approximation it is assumed

$$F(\sigma) = \sqrt{\frac{2}{\pi}}\frac{\delta_c}{\sigma}\exp\left(-\frac{\delta_c^2}{2\sigma^2}\right), \tag{4}$$

with $\delta_c = 1.69$. However, recently it has been shown that

$$F(\sigma) = A\exp(-|\ln(1/\sigma) + B|^\varepsilon) \tag{5}$$

provides a better fit to halo mass functions obtained from large N-body simulations (Jenkins et al., 2001; Evrard et al., 200; Hu & Kravtsov, 2003). The parameters A, B and ε seem to be independent of cosmology and redshift when halo masses are defined with respect to the background density. In Jenkins et al. (2001) it was obtained $A = 0.315$, $B = 0.61$ and $\varepsilon = 3.8$, with the halo mass being defined as that enclosed within a region overdense by a factor of 180 with respect to the background density. In Evrard et al. (200) the halo mass was instead defined as that which corresponds to an halo overdensity of 200 with respect to the critical density, leading to $A = 0.22$, $B = 0.73$ and $\varepsilon = 3.86$ for the case of a $\Omega_0 = 0.3$ ΛCDM flat cosmology. Note that in the Press-Schechter approximation the halo mass is that given by the virial relation (Bryan & Norman, 1998).

From the previous expressions one finds that $n(M, z)$ depends only on σ_8, Ω_0, Ω_w, w, and Γ (in decreasing order of importance). Given that $n(M, z)$ varies with M and z differently for distinct combinations of these parameters, in principle knowing how $n(M, z)$ changes can lead to the estimation of any of those parameters. Observationally, the most accessible interesting quantity is the number density of rich clusters of galaxies at the present epoch, providing an estimate for $n(M, z \sim 0)$ in the range of scales 10^{14} to $2 \times 10^{15} h^{-1} M_\odot$. Given that $n(M, z)$ depends most strongly on σ_8, traditionally the present-day abundance of galaxy clusters has been used to determine this parameter (as a function of the others). Data on the evolution with redshift of the cluster number density has in turn been used to try to break the degeneracy between σ_8 and the next most influential parameter, Ω_0, simultaneously estimating both (again as a function of the remaining parameters).

3. Using the local cluster abundance to constrain σ_8

The best method to find clusters of galaxies is through their X-ray emission, which is much less prone to projection effects than optical identification. The selection function of X-ray cluster catalogues is therefore usually well characterized, which is essential if the cluster abundance, and hence σ_8, is to be determined with small uncertainty. Further, the X-ray temperature of a galaxy cluster is at present the most reliable estimator of its mass, thereby allowing the correct comparison between the observed cluster abundance and the halo abundance predicted by structure formation models. It is then not surprising that most determinations of σ_8 using the number density of rich clusters of galaxies at the present epoch have relied on X-ray selected galaxy cluster catalogues (see Viana, Nichol & Liddle, 2002; and Viana et al., 2002; for references).

Traditionally, the relation between cluster mass and X-ray temperature has been determined using N-body hydrodynamical simulations. To a large extent this was due to the limitations of past X-ray satellites, which did not possess enough spatial resolution for the cluster masses to be well constrained. Although the situation has changed significantly with the launch of the CHANDRA and XMM-NEWTON satellites, there is still the problem that X-ray data with good enough quality to allow the determination of the cluster mass profile over most of the cluster radius is not yet available (see. e.g. Pratt & Arnaud, 2002). Further, such calculation relies on the hydrostatic equilibrium hypothesis, which may not be a good approximation for a significant fraction of the galaxy clusters (see Smith et al., 2002; and references therein). Thus, for now, N-body hydrodynamical simulations seem to remain the best means to determine the cluster mass to X-ray temperature relation over most of the cluster radius. However, it has recently become apparent that traditional hydrodynamical simulations, where the gas is only allowed to heat adiabatically and through shocks, have difficulties in matching observations in the central regions of clusters, with a significant underestimation of the temperature corresponding to a given cluster mass. In a recent paper (Thomas et al., 2002) it has been showed that the inclusion of extra gas physics, namely radiative cooling of the gas and its preheating before cluster formation, can bring simulations into good agreement with recent CHANDRA observations of the cores of clusters (Allen, Schmidt & Fabian, 2001), suggesting that these may be crucial ingredients in obtaining an accurate description of clusters.

In the light of these new results on the cluster mass to X-ray temperature relation (which also have the advantage of providing an estimate for the scatter in the relation), and recently available better quality data on the present-day cluster abundance (Reiprich & Böhringer, 2002; Ikebe Y. et al., 2002), we decided to re-estimate σ_8 using a novel Monte Carlo based approach (Viana et al., 2002).

It is beyond present computational means to directly extract X-ray cluster catalogues from hydrodynamical simulations due to the excessive number of particles required to obtain statistically-robust cluster abundances with temperatures above a few keV. Instead, we appeal to the method used in Holder, Haiman & Mohr (2001), which is to use generalized mass functions of dark matter haloes (in our case as given in Evrard et al., 200), to generate catalogues of clusters identified by their redshift and mass, and then estimate their X-ray temperatures using the cluster mass to X-ray temperature relation obtained in hydrodynamical simulations. Here we use the results of those described in (Thomas et al., 2002, Muanwong et al., 2002). Given that these simulations have only been carried out for the currently-favoured ΛCDM cosmological model, with $\Omega_0 = 0.35$ and $\Omega_\Lambda = 0.65$, we restrict the estimation of σ_8 to this cosmology. We assume $\Gamma = 0.18$, which is the favoured value from the joint analysis of the 2dF (Percival et al., 2001) and SDSS (Szalay et al., 2001; Dodelson et al., 2002) data, when accounting for both statistical and systematic uncertainties (the allowed interval for Γ is [0.08, 0.28] and we

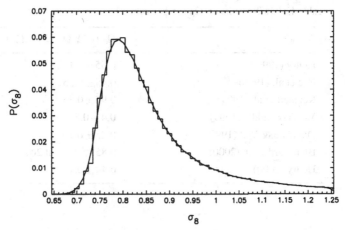

Figure 1. Marginalized probability distribution for σ_8.

confirm that varying Γ within this interval leads to almost no change in the final results).

In order to compare the simulated catalogues with the data we need to impose the same flux selection criterion as used in defining the observed cluster sample, which forces us to use a relation between X-ray luminosity (in the [0.1, 2.4] keV band) and temperature. In order to be coherent, we determine this relation from the data simultaneously with σ_8. We take it to be a power-law of the form

$$\log_{10}(L_X/h^{-2}\,\mathrm{erg\,s^{-1}}) = A + \alpha\log_{10}(kT/\mathrm{keV}), \qquad (6)$$

with a dispersion $\sigma_{\log_{10} L_X}$ assumed to be independent of temperature. The set of 4 free parameters considered most correct will be the one that not only most often reproduces the observed number of clusters but also most often closely reproduces the observed distribution of the cluster properties (z, kT, L_X).

The marginalized probability distribution for σ_8 over the 3 others is presented in Figure 1. We obtain $\sigma_8 \simeq 0.79$ within [0.73, 1.15] at the 95 per cent confidence level, in agreement with the results of Ikebe Y. et al. (2002) and Pierpaoli et al. (2002). Also, $A \simeq 42.0$, $\alpha \simeq 2.7$ and $\sigma_{\log_{10} L_X} \simeq 0.25$, are the best-fit values when marginalized over the others.

Our results do not indicate a dramatic reduction in σ_8 derived from the abundance of X-ray clusters, as has been recently claimed (Borgani et al., 2001; Seljak, 2001; Smith et al., 2002), and support the weak lensing measurements of σ_8 that put it around 0.8 for the cosmology we consider here (Bacon et al., 2002; Hökstra, Yee & Gladders, 2002; Refregier, Rhodes & Groth, 2002; Van Waerbeke et al., 2002), though they are also in most cases perfectly compatible within the uncertainties with the weak lensing estimates that prefer lower values (Brown et al., 2002; Hamana et al., 2002; Jarvis et al., 2002).

Table I. *EMSS* results on Ω_0 (95 per cent c.l.)

Author	Ω_0 (with $\Omega_\Lambda = 1 - \Omega_0$)
Henry (1997)	0.55 ± 0.17
Eke et al. (1998)	0.45 ± 0.25
Reichart et al. (1999)	0.95 ± 0.35
Viana & Liddle (1999)	$0.4 - 0.8$
Donahue & Voit (1999)	0.25 ± 0.10
Blanchard et al. (2000)	$0.85 \; (> 0.35 \text{ at } 2\sigma)$
Henry (2000)	0.44 ± 0.12

4. Using the cluster abundance evolution with z to estimate Ω_0

We have already mentioned that Ω_0 can in principle be simultaneously estimated together with σ_8, as long as cluster abundance data for different
redshifts is available. Ideally, this data would refer to a wide range of redshifts, with X-ray temperatures determined for all clusters. However, we still
do not have such a catalogue available, with the closest being the *EMSS*
(Gioia et al., 1990; Henry et al., 1992; Gioia & Luppino, 1994; Lewis et al.,
2002), *RDCS* (Rosati et al., 1998; Holden et al., 2002) and *Bright-SHARC*
(Romer et al., 2000; Adami et al., 2000). The first has been by far the most
extensively used catalogue for the propose we are discussing. It has the advantage over the others of having had the X-ray temperature measured for all its
clusters (tough that has been almost done also for the *Bright-SHARC*), besides
the fact that among the 3 catalogues mentioned it is the one with the largest
sky coverage (followed by the *Bright-SHARC* and then by the *RDCS*). The
RDCS and *Bright-SHARC* catalogues are however interesting in their own
way: the *RDCS* is the deepest, with a flux selection criterion about 10 times
lower than either the *EMSS* or the *Bright-SHARC*; while the *Bright-SHARC*
is that for which the catalogue selection function has been studied in most
detail. Ideally, we would like to bring together the strengths of each of these
catalogues: the large sky coverage of the *EMSS*; the sensitivity of the *RDCS*;
the thoroughness with which the *Bright-SHARC* selection function was studied. This could be achieved with an all-sky survey obtained with a new X-ray
satellite, but unfortunately it is far from certain that there will be one in the
near future. The next best thing may well be the *XCS* (Romer et al., 2001).
Meanwhile, we summarize in Table 1 all the estimates of Ω_0 that have been
obtained based on the *EMSS* catalogue.

The large dispersion in the results is to a great extent due to the limitations
of the data, which means that the results are very sensitive to differences in
the various analysis. The *RDCS* catalogue has only been analyzed in Bor-

gani et al. (1999) and Borgani et al. (2001), who obtained $\Omega_0 \simeq 0.25$ (within [0.1, 0.4] at 95 per cent confidence). The analysis of the *Bright-SHARC* catalogue is currently under way (Viana et al., 2003), with preliminary results indicating $\Omega_0 \sim 0.35$.

5. The future is the XCS

We have just concluded that there is a pressing need for a new galaxy cluster catalogue, of greater size, and in particular going to higher redshift, than existing ones, and with a well understood selection function. In (Romer et al., 2001) we describe in considerable detail how such a catalogue may be constructed through serendipitous detections of galaxy clusters in archival data from the XMM-NEWTON satellite. By examining the many thousands of pointings which will be made, it will be possible to build a representative sample of randomly, and hence objectively, selected X-ray clusters. The *X-ray Cluster Survey (XCS)* will not only be an invaluable resource for cosmological studies, but will also have a variety of other applications (Romer et al., 2001). Here, we briefly describe the calculation, by means of a likelihood analysis, of the uncertainty associated with the estimation of cosmological parameters using the *XCS*.

In order to determine how many, and what type of, clusters the *XCS* might detect, we have calculated in (Romer et al., 2001) the survey sensitivity limit as a function of several parameters, including cluster X-ray temperature and redshift, exposure time, telescope vignetting, and cosmological parameters, which was then folded with the expectation on the number density of clusters as a function of X-ray temperature and redshift for the assumed cosmological parameters. Further, it was also determined what fraction of these clusters would have their X-ray temperatures reliably estimated solely from the serendipitous data, which normally requires a minimum of 1000 photons (Liddle et al., 2001). The survey sensitivity limits used in the likelihood analysis were calculated in the same manner as in (Romer et al., 2001), but slightly differ in that we used the actual in-orbit particle background instead of that which was predicted pre-launch (Romer et al., 2002).

The likelihood calculation was performed by means of a Monte Carlo method similar to that used in (Holder, Haiman & Mohr, 2001), whereby 1000 realizations of the expected *XCS* catalogue were generated for an input fiducial cosmological model, with the addition of Poisson noise, and then for each realization the cosmological parameters σ_8, Ω_0 and Ω_Λ, were allowed to vary so as to find their most probable values given each catalogue. The input fiducial model was the currently-favoured spatially-flat low-density cosmology with $\Omega_0 = 0.3$ and $\Omega_\Lambda = 0.7$, for which we assumed $\sigma_8 = 0.8$. In it, structure formation is assumed to proceed through gravitational instability from

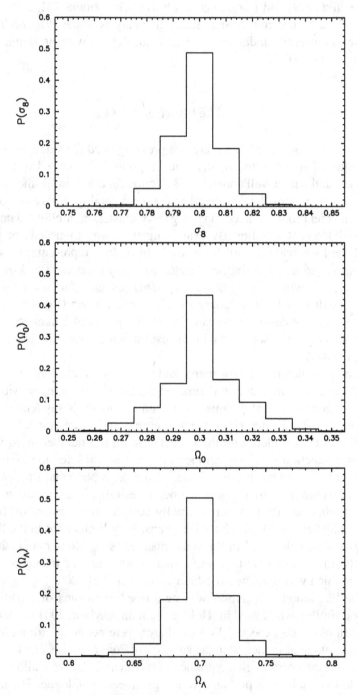

Figure 2. Probability distribution for the three parameters, σ_8 (top), Ω_0 (middle) and Ω_Λ (bottom), we want to constrain with the *XCS*, given the input fiducial model.

a Gaussian distribution of primordial density perturbations, and the current shape of the linear power spectrum of these perturbations is taken to be of the kind expected in CDM models, being the shape parameter fixed at $\Gamma = 0.2$.

The expected number of *XCS* clusters with measurable X-ray temperatures (above 2 keV), as a function of cosmological parameters, was computed by combining the *XCS* selection function with the mass function obtained through large N-body simulations (Jenkins et al., 2001) and a cluster X-ray temperature to mass relation based on data from hydrodynamical N-body simulations (Gus Evrard, private communication).

In Figure 1 we show the histograms representing the probability distributions for the three parameters σ_8 (top), Ω_0 (middle), and Ω_Λ (bottom), given the assumed underlying model.

The conclusion is that the *XCS* will provide competitive estimates for the three most important cosmological parameters, σ_8, Ω_0, and Ω_Λ, enabling their joint estimation to within respectively 3, 10 and 7 per cent of their true values at the 95 per cent confidence level. Of course, the uncertainties would be even smaller if the number of unknown parameters was reduced, for example if flatness was assumed: $\Omega_0 + \Omega_\Lambda = 1$. However, it could also be argued that the number of unknown parameters should be increased by the introduction of extra ones describing the equation of state of the dark energy component.

Acknowledgements

I would like to thank all my collaborators, whose contribution to the work presented here has been essential. And, in particular, Andrew Liddle, Kathy Romer and Bob Nichol, for their encouragement.

References

Adami, C., et al.: 2000, *ApJS* **131**, 391

Allen, S.W., Schmidt, R.S., Fabian, A.C.: 2001, *MNRAS* **328**, L37

Bacon, D., Massey, R., Refregier, A., Ellis, R.: 2002, *MNRAS* , submitted [astro-ph/0203134]

Blanchard, A., Sadat, R., Bartlett, J.G., le Dour, M.: 2000, *A&A* **362**, 809

Borgani, S., Rosati, P., Tozzi, P., Norman, C.: 1999, *ApJ* **517**, 40

Borgani, S., et al.: 2001, *ApJ* **561**, 13

Brown, M.L., et al.: 2002, *MNRAS* **341**, 100

Bryan, G.L., Norman, M.L.: 1998, *ApJ* **495**, 80

Dodelson, S., et al. (the SDSS collaboration): 2002, *ApJ* **572**, 140

Donahue, M., Voit, G.M.: 1999, *ApJ* **523**, L137

Eke, V.R., Cole, S., Frenk, C.S., Henry, J.P.: 1998, *MNRAS* **298**, 1145

Evrard, A.E., et al.: 2002, *ApJ* **573**, 7

Gioia, I.M., et al.: 1990, *ApJS* **72**, 567

Gioia, I.M., Luppino, G.A.: 1994, *ApJS* **94**, 583

118   Pedro T. P. Viana

Hamana, T., et al.: 2002, *ApJ* , submitted [astro-ph/0210450]
Henry, J.P., et al.: 1992, *ApJ* **386**, 408
Henry, J.P.: 1997, *ApJ* **489**, L1
Henry, J.P.: 2000, *ApJ* **34**, 565
Hökstra, H., Yee, H.K.C., Gladders, M.D.: 2002, *ApJ* **577**, 595
Holden, B.P., et al.: 2002, *AJ* **124**, 33
Holder, G., Haiman, Z., Mohr, J.J.: 2001, *ApJ* **560**, L111
Hu, W., Kravtsov, A.V.: 2003, *ApJ* **584**, 702
Ikebe, Y., Reiprich, T.H., Böhringer, H., Tanaka, Y., Kitayama, T.: 2002, *A&A* **383**, 773
Inoue, K.T., Nagashima, M.: 2002, *ApJ* **574**, 9
Jarvis, M., et al.: 2002, *AJ* **125**, 1014
Jenkins, A.R., et al.: 2001, *MNRAS* **321**, 372
Lewis, A.D., Stocke, J.T., Ellingson, E., Gaidos, E.J.: 2002, *ApJ* **566**, 744
Liddle, A.R., Viana, P.T.P., Romer, A.K., Mann, R.G.: 2001, *MNRAS* **325**, 875
Muanwong, O., Thomas, P.A., Kay, S.T., Pearce, F.R.: 2002, *MNRAS* **336**, 527
Percival, W.J., et al. (the 2dF team): 2001, *MNRAS* **327**, 1297
Pierpaoli, E., Borgani, S., Scott, D., White, M.: 2002, *MNRAS* , submitted [astro-ph/0210567]
Pratt, G.W., Arnaud, M.: 2002, *A&A* **394**, 375
Press, W.H., Schechter, P.: 1974, *ApJ* **187**, 425
Refregier, A., Rhodes, J., Groth, E.J.: 2002, *ApJ* **572**, L131
Reichart, D.E., et al.: 1999, *ApJ* **518**, 521
Reiprich, T.H., Böhringer, H.: 2002, *ApJ* **567**, 716
Romer, A.K., et al.: 2000, *ApJS* **126**: 209
Romer, A.K., Viana, P.T.P., Liddle, A.R., Mann, R.G.: 2001, *ApJ* **547**, 594
Romer, A.K., et al.: 2002, in *Tracing Cosmic Evolution with Galaxy Clusters*, Proceedings of the Sesto-2001 Workshop, (eds) S. Borgani, M. Mezzetti and R. Valdarnini, *A.S.P. Conf. Ser.* **268**, 43
Rosati, P., della Cecca, R., Norman, C., Giacconi, R.: 1998, *ApJ* **492**, L21
Seljak, U.: 2001, *MNRAS* **337**, 769
Smith, G.P., et al.: 2002, *ApJ* , **590**, L79
Szalay, A.S., et al. (the SDSS collaboration): 2001, *ApJ* , submitted [astro-ph/010741]
Thomas, P.A., Muanwong, O., Kay, S.T., Liddle, A.R.: 2002, *MNRAS* **330**, L48
Van Waerbeke, L., et al.: 2002, *A&A* **393**, 3690211090
Viana, P.T.P., Liddle, A.R.: 1996, *MNRAS* **281**, 323
Viana, P.T.P., Liddle, A.R.: 1999, *MNRAS* **303**, 535
Viana, P.T.P., Nichol, R.C., Liddle, A.R.: 2002, *ApJ* **569**, L75
Viana, P.T.P., Kay, S.T., Liddle, A.R., Muanwong, O., Thomas, P.A.: 2002, *MNRAS* , submitted [astro-ph/0211090]
Viana, P.T.P., Nichol, R.C., Liddle, A.R., Romer, A.K., Adami, C., Ulmer, M.P., Collins, C.A., Burke, D.J.: 2003, in preparation
White, M.: 2002, *ApJS* **143**, 241

The VLTI: Challenges for the Future

Francesco Paresce

European Southern Observatory, D-85748 Garching b. Muenchen, Germany

2003 January 22

Abstract. ESO's VLT Interferometer is quickly ramping up to its full potential as the first instruments covering the near and mid IR are deployed and the 8m telescopes are equipped with state-of-the-art adaptive optics systems. I will review the expected scientific benefits of such improvements and outline the next steps in the continuing quest for higher spatial resolution and sensitivity.

Keywords:

1. Introduction

Figure 1 shows graphically the real challenge for the VLT Interferometer (VLTI). On the left, the best currently available 2D image taken by an optical interferometer is represented while, on the right, is shown a typical 2D image taken routinely nowadays by a radio interferometer.

The comparison is striking as it shows better than the proverbial one thousand words the long road optical long baseline interferometry (OLBIN) has to go before it can rival its cousin and reach the scientific objectives it has

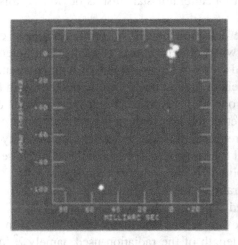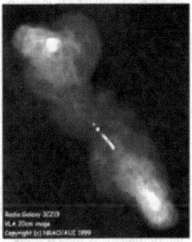

Figure 1. The triple star Eta Virginis observed in the optical with the Navy Prototype Optical Interferometer's 6-station array (left panel) and the radio galaxy 3C219 observed with the VLA at 20 cm (right panel).

set for itself. The latter, of course, include obtaining pictures in the optical and near and mid infrared of astronomical targets such as planets, stellar accretion disks and jets, supernovae ejecta and the nuclei of nearby AGN with the resolution, clarity and fidelity of the type shown in the right panel of Figure 1.

Going from left to right in this figure is no easy task from the technical standpoint for a number of reasons that I will describe in the next paragraphs but some outstanding results obtained recently such as the one shown in the left panel of Figure 1 raise the distinct possibility that OLBIN will soon be able to fulfill its promise and we may soon see a period coming where images of the caliber routinely obtained in the radio are available in the optical at similar if not better spatial resolutions. I will use the case of the VLTI, the only currently operating very large telescope array designed from the start as an interferometer, as an example of the type of OLBIN facility that will carry the burden of transforming the potential into reality by the end of this decade.

2. Radio Versus Optical Long Baseline Interferometry

In order to understand the basic reasons behind the evident lag in performance between the two techniques let us consider some fundamental facts. In the radio range (10-600 THz), the atmospheric coherence time is typically of order $\sim$20 minutes and the isoplanatic patch $\sim$1-$2°$ on the sky while the same parameters appropriate to the optical are typically 1-10 ms and 15-30'. This clearly means that high precision phase referencing in the radio range can be done by simply off-pointing to a calibrator star. Just as obviously, this is simply impossible in the optical.

Moreover, the physics of the detection process in the radio allows one to use heterodyne combination wherein detection is carried out at antenna level, the signal amplified, mixed with a reference signal of high coherence, digitized and combined in a correlator later at one's leisure. In contrast in the optical, this is not possible and one must resort to homodyne combination. In this case, one must directly detect the interference pattern only after correction for the optical path difference (OPD) and, since only the intensity of the field can be reliably detected the fringes must be modulated in phase to measure both the phase and amplitude of the complex visibility.

Some other practical considerations that have limited up to now the usefulness of OLBIN are associated with the precision and sensitivity of the combination process itself. For example, the maximum OPD allowed in an interferometer is the coherence length of the radiation used namely $\lambda_0^2/\Delta\lambda$ where λ_0 and $\Delta\lambda$ are the central wavelength and bandpass of the device, respectively which, in the optical at $1\,\mu$, corresponds to $\sim 10\,\mu$ over the baseline of $\sim$200m or more or $\sim$ a part in 10^9. If that were not challenging enough, the optical path jitter during the observation must not smear the fringes i.e. the dynamical stability of the OPD compensation must be of order of 10nm. No wonder the technique has been so difficult to apply until recently.

The situation on the sensitivity was equally bleak up until the availability of accurate fringe trackers and very large telescopes equipped with adaptive optics systems. To understand this problem, consider the fact that the signal to noise ratio S/N in an optical interferometer is proportional to NV^2 where N corresponds to the number of photons detected per subaperture and per integration time and V is the measured visibility of the resulting fringe pattern. Thus, paradoxically, the detection accuracy decreases significantly with increasing source complexity in the sense that the correlated magnitude of a resolved object m_{corr} is related to its unresolved magnitude m by $m_{corr} = m - 2.5 \log V^2$. This implies, for example, that an object of magnitude m whose visibility at a certain baseline is 10% (i.e. that the ratio of coherent to total radiation is 0.1) will have a correlated magnitude $m_{corr} = m+5$ or, in other words, a sensitivity 100 times less.

Considering, finally, the twin facts that, without a fringe tracker, integration times are limited to $\sim$10ms at best, that the background due to detector noise, sky and thermal emission from the telescope is high (especially at $10\,\mu$), an optical interferometer's sensitivity or limiting magnitude is shockingly low. Even with telescopes of $\sim$1m in diameter, the limiting uncorrelated magnitude of typical current interferometers is only K $\sim$ 5-7 in the best conditions.

3. Fundamental OLBIN Requirements

So, what do we need to reach our objectives of 2D imaging in the optical/IR at 1mas resolution and K=20 sensitivity that are both a must for astronomical competitiveness in the next few years?

We need an efficient array of the largest possible number n of telescopes with diameter D. The number of independent baselines increases rapidly with n (as n(n-1)) and, with it, the u,v plane coverage that is essential for accurate reconstruction of complex sources. This requirement is best shown in Figure 2 where the effect of u,v plane coverage on the shape of the synthesized beam is illustrated clearly. Starting from the situation shown in the top right where the sparse coverage of the u,v plane corresponds to a very dirty and essentially unusable beam, progressing down to the bottom right where the situation gets considerably better mainly due to the effect of earth rotation synthesis and, finally, to the left where a combination of more and/or mobile telescopes and earth rotation synthesis provides the ultimate resolution in a compact, dominating core of milliarcsecond (mas) size using a baseline of $\sim$100m projected length.

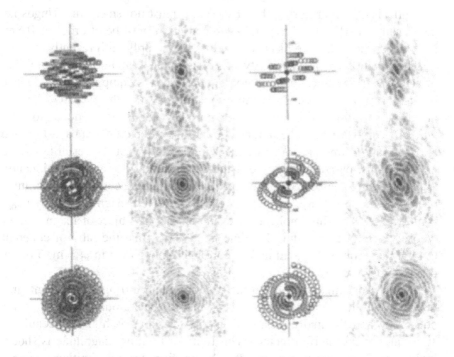

Figure 2. The effect of u,v plane coverage on the synthesized beam. Each panel consists in the u,v points corresponding to the particular configuration considered on the left and the corresponding synthesized beam on the right. Coverage increases from the top right to the bottom left.

We also need the largest possible baselines lengths B coupled with the largest possible D to maximize resolution and sensitivity. All telescopes of D>~1m need to be AO equipped to be used to their full potential. Extremely stable delay lines (DL) and accurate fringe trackers that allow the fringe packet to be frozen and, therefore, to allow integration times to grow hundred fold are other essential desiderata. n-way (with n≥4) beam combiners covering the optical/IR range with spectral resolution of up to 10^4 will also have to play an important role to overcome the inherent limitations of the technique brought about by the effects of the turbulent atmosphere by allowing the efficient use of standard closure phase techniques. Finally, the ultimate performance of OLBIN for faint object imaging and high precision astrometry will be attained when dual feed systems permitting the use of bright stars as phase reference for much fainter targets will be developed in the near future.

4. Where Are We? Current Status of VLTI

How, then, does the current instrumentation compare to these requirements? The situation for what regards the VLTI at the end of 2002 is that all four very large telescopes (UT1-4) have been integrated into the VLTI array in the configuration shown in Figure 3. All four telescopes have been combined in pairs to observe a number of targets successfully. Five baselines have been used now out of the available six, UT3/UT4 being the only one not used yet as there is only one delay line on the required side of the tunnel.

In particular, the UT1 and 3 telescopes have been equipped with tip/tilt sensors and used extensively to commission the array and obtain preliminary science results with the combination's 103m baseline. In addition, two 40cm diameter siderostats in the positions indicated in Figure 3 with 16, 66 and 140m long baselines were also used as the workhorses for system verification and are being used to carry out shared risk observations of astronomical targets proposed by community astronomers.

Currently, three 60m long-stroke delay lines are operating in the 120m-long tunnel at the center of the observation platform of Cerro Paranal shown in the left panel of Figure 3. These delay lines are the system's key feature that allows it to reach the stringent beam compensation and tracking requirements alluded to earlier. The measured flatness of the rails is now better than $25\,\mu$m over 65m with an absolute position accuracy of $30\,\mu$m. The relative position error of the carriages has been measured to be $\sim$20nm over a 50ms integration

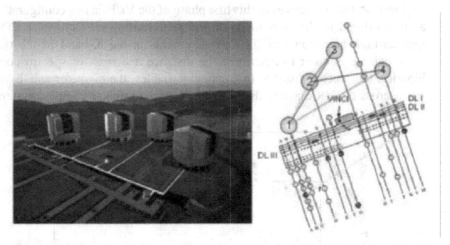

Figure 3. The observing platform on Cerro Paranal with the 4 UTs, the 120m long delay line tunnel and the beam combination laboratory at the center of the VLTI array where the beams from the various telescopes converge and the stations for the auxiliary telescopes (left panel). The schematic configuration of the same array is shown on the right panel. The lines indicate the baselines used so far between the four UTs and the two siderostats located on the AT stations. DL I,II,III correspond to the 3 delay lines.

time. The beams are combined in the pupil plane with the test instrument VINCI operating in the K band.

The available facility is described in much greater detail in the proceedings of the conference *Interferometry for Optical Astronomy II*, (ed.) Traub, W., Proc. SPIE 4838, 2003 and in the relevant ESO website:

http://www.eso.org/projects/vlti/

To appreciate how well this facility is working in practice, Figure 4 shows the instrument's transfer function obtained with observations of Fomalhaut used both as the target and calibrator in a series of consecutive measurements with VINCI during a whole night in October, 2002 with the siderostats and the 140m baseline (Davis et al. 2003, submitted to A&A). The transfer function is defined as the ratio of the observed visibility squared to the expected visibility squared. The true visibility squared for a given object is then obtained by dividing its observed visibility squared by the transfer function for its associated calibrator.

As can be seen from this figure no significant variation in the transfer function was found for the zenith angle range 5° to 70°. The projected baseline varied between 139.7m and 49.8m during the observations and, as an integral part of the determination of the transfer function, a new accurate limb-darkened angular diameter for Fomalhaut of 2.112 ± 0.011 mas has been established. This has led to improved values for the emergent flux of $3.42 \pm 0.10 \times 10^8$ Wm^{-2}, an effective temperature of 8813 ± 66 K, a radius of $1.214 \pm 0.011 \times 10^9$ m (R/R$_\odot$ = 1.746 ± 0.016) and a luminosity of $6.33 \pm 0.22 \times 10^{27}$ W (L/L$_\odot$ = 16.5 ± 0.6).

The scientific objectives of this first phase of the VLTI in this configuration as it is offered to the community in the year starting on October 1, 2002 include high precision visibility measurements in the K band of the type shown above in order to determine the distance and mass of spectroscopic binaries, the precise distance of Cepheid variables, the diameter and T$_{eff}$ of stars across the HR diagram and a first look at the brighter complex systems

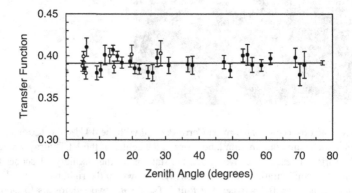

Figure 4. The transfer function of the VLTI as a function of zenith angle.

like pre-main sequence stellar and debris disks and the envelopes and shells of evolved stars. In general, the observational philosophy in this first phase where we are restricted to measuring visibilities and not phases and, therefore, to objects having an axially symmetric intensity distribution is to obtain several hundred measurements at a 1% precision or better rather than several thousand measurements at 10% precision that would leave theoretical models unconstrained.

5. VLTI: Immediate Prospects (>2003)

Two new instruments will become available in the very near future namely in the period 2003-2004. They will extend the capabilities of the VLTI in three very significant ways: extended wavelength coverage, 3 way beam combination allowing phase closure imaging and increased spectral resolution. One of these is MIDI that will combine two beams in the N band (8-12 μ) with a spectral resolution of 100. Its limiting magnitude (uncorrelated) is expected to reach N=5 with the UTs with a field of view in the range 0.26-1.14$''$ and a minimum fringe spacing of 10 mas. MIDI obtained first fringes with the UTs in December, 2002 (see http://www.eso.org/outreach/press-rel/pr-2002/pr-25-02.html). The other will be AMBER operating in the J,H,K$'$ bands (1-2.4 μ) at selectable spectral resolutions of 35, 1000, and 10,000. Its field of view will be in the range 0.06-0.24$''$ with a minimum fringe spacing of 1-2 mas and reaching a limiting uncorrelated magnitude of K=13 with the UTs. The exceptional aspect of this instrument will be its ability to combine 3 beams simultaneously.

But much more is on the way in this period. The FINITO fringe tracker will allow on axis fringe tracking in the H band of bright sources (H<12) observed at longer wavelengths thus significantly increasing the visibility measurement accuracy of these objects and the limiting magnitude of both instruments. Of similar importance will be the arrival of two adaptive optics systems on two UTs at the second half of 2003. This will allow removing almost all aberrations except piston with a Strehl ratio of 50% in K for a V<13 reference star and 25% for a V<16 star. The expected increase in limiting magnitude will be considerable. Two more AO systems will be added to the remaining UTs by 2004. Three more delay lines will also be installed in the next year allowing coverage of more than 90% of the AT stations.

Finally, the first two 1.8m moveable ATs will be integrated in the VLTI facility by mid-2004 and account for an enormous increase in the critical u,v plane coverage as discussed in the previous section. They are already now at an advanced stage of construction as shown in Figure 5.

They represent a special plus for the facility in that they can be used exclusively for interferometry without having to resort to the already heavily subscribed UTs. Of course, all this equipment will create a serious traffic jam at the combination point as shown in Figure 6.

The scientific objectives of this second phase of the VLTI development beyond ∼ 2003 are all centered on the exploitation of the increased sensitivity and precision of the facility with the various devices described above. In particular, the capability to perform real i.e. model-independent imaging on moderately bright complex sources (K<14) by determining the phase of the fringe packet in addition to its visibility opens up the possibility of accurate and faithful reconstruction of astronomical scenes. These would include, for example, the morphology of dust torii in nearby AGN, the structure of the circumstellar envelopes of mass losing giants, and long lived stellar surface features such as spots and faculae in the star's magnetic network.

Closure phase techniques with the 3-way beam combiner in AMBER and the mid IR capability of MIDI will be very useful in studying and, hopefully, understanding the complex and currently uncertain relationship between planets and the stellar accretion and debris disks from which they originate. The size, temperature structure and possible features in these disks due to the presence of protoplanets should be accurately measurable especially, possibly for the stars that are already known to possess planets via the radial velocity technique. Already the measurement of the binary fraction in nearby clusters and star forming regions will go a long way in elucidating the basic mechanisms underlying the complex process of star formation.

6. VLTI: Far Future Prospects (> 2005)

There are two effective ways to measure both the visibility and phase of the fringe packet formed at the beam combination point of the array. The first is closure phase that requires n-way beam combiners where n≥3. The larger n

Figure 5. The AT being assembled in Europe (leftmost panel) and two views of the transporter completely assembled (right panels).

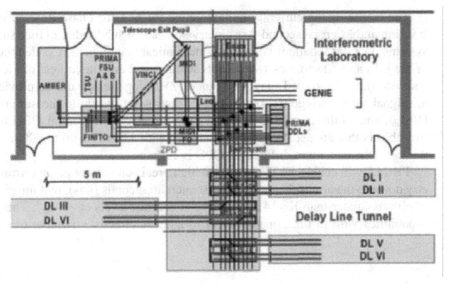

Figure 6. Beam location and distribution at the beam combination lab at the center of the array. ZPD indicates the Zero Path Difference position while DL is a delay line. The beam compressors convert the beam sizes from the various feeding telescopes to a standard value required for the instruments.

the better, of course, but the technique is restricted for the moment to relatively bright and stable objects of $K < 14$ at best with the UTs. A better way is to use phase referencing using a bright reference star in the field of view in the fashion of adaptive optics. This allows high fidelity image reconstruction of faint complex sources ($K < 20$ with the UTs) situated close to a bright reference. Close in the K band means within a radius of $30'$ approximately. The way this works is indicated schematically in Figure 7.

Controlling all optical path lengths of the reference star and of the science star inside the interferometer (OPD_{int}) with a laser metrology system introduces the capability of imaging faint objects and of determining the precise angular separation between the two stars. The measurement has to be repeated for up to 30 min in order to average out the variations of the differential OPD caused by atmospheric turbulence (OPD_{turb}). With the two OPD terms being determined, the measurable is the sum $\Delta SB + \phi$. If both stars are point-like, the phase ϕ of the visibility function is zero, and if the baseline B is known with high accuracy, one obtains a high precision astrometric measurement of the angular separation ΔS. If only the reference star is point-like and the science object is an extended object with a non-symmetric structure, the (non-zero) phase ϕ of the visibility function depends on the baseline vector B. Then, the measured sum $\Delta SB + \phi$ can be disentangled by repeating the measurement for several different baselines.

The facility being currently built is called PRIMA for Phase Referenced Imaging and Microarcsecond Astrometry and can be subdivided in four subsystems: 1) a star separator i.e. an opto-mechanical system in the Coudé focus of the UTs and ATs to pick two objects within the 2 arcminute field of view and send the light to the Delay Line tunnel, 2) a fringe sensor unit to provide the signal for the fringe tracker, 3) a laser metrology system to measure the OPD_{int}, and 4) differential Delay Lines to correct the differential OPD for two objects that are separated up to 2 arcmin with a baseline up to 200m. In this case, the maximum differential OPD is 130mm.

PRIMA will enable VLTI to perform high precision narrow-angle astrometry down to the atmospheric limit of 10 microarcseconds (μas), real imaging of objects fainter than K$\sim$14 and nulling at contrast levels of $\sim 10^{-4}$. These capabilities will, in turn, enable VLTI to address directly a number of ex-

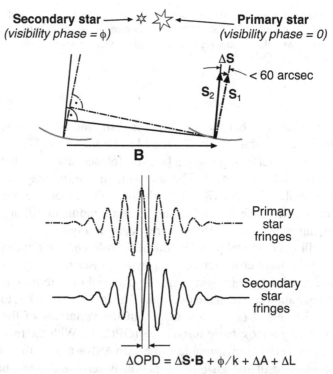

Figure 7. Schematic of the process of phase referenced imaging and astrometry. The stabilization of the fringe packet of the faint science object (secondary star, solid line) is accomplished by using a bright guide star (primary star, dot-dash line) as a reference. The difference in the positions of the two fringe packets are determined by the OPD given by the product of ΔS – the angular separation vector of the stars – and B – the baseline vector – by the phase ϕ of the visibility function of the science object, and by the OPD caused by the atmospheric turbulence and by the internal OPD.

tremely important scientific issues that are currently at the top of the list of challenges for future astronomical high resolution instrumentation.

The first is the detection and characterization of extra-solar planets and their birth environment. This objective requires application of both the astrometric and the imaging capability of PRIMA. The astrometric capability will determine the main physical parameters of planets orbiting around nearby stars already found by radial velocity (RV) techniques including their precise mass, orbital inclination and a low resolution spectrum.

In addition and most importantly, astrometry will allow VLTI to extend the search for planets to stars that cannot be properly covered by the RV technique, in particular pre-main sequence (PMS) stars. The crucial exploration of the initial conditions for planetary formation in the stellar accretion disk as a function of age and composition will be finally possible. The simultaneous image at mas resolution or better of the accretion disk from which any particular planet is born will add a new and exciting dimension to our understanding of both planetary and stellar formation mechanisms since the complex accretion disk is expected to be the cradle of these objects.

Another key objective for VLTI with PRIMA will be the exploration of the nuclear regions of galaxies including our own. The resolution of the VLTI at 2μ corresponds to 15 AU at the galactic center or about 1500 times the Schwarzschild radius of a 10^6 $M_\odot$ black hole (BH). The first and most important goal will be to test for the presence of a central massive BH by measuring the three-dimensional velocity field of the star cluster centered on IRS16. The VLTI will be able to probe the 3D space motions of stars in the nuclear cluster down to 10^{-4} pc of the central source or approximately two orders of magnitude better than currently. This data will certainly provide very precise information on the central mass without the a-priori assumption of isotropic motions. High precision astrometry with PRIMA might even go so far as to probe the central BH to a few times its Schwarzschild radius (2.5×10^{-7} pc).

At this point, one can finally say that OLBIN will have reached the major leagues and be able to compete mas per mas with the radio facilities like the VLA and images such as the one shown on the right panel of Figure 1 obtained routinely in the near O/IR.

The VLTI as presently designed can accomplish this ambitious but absolutely critical task by relying on its capability to support high precision phase referencing. This characteristic will be critical in order to perform the high precision, narrow angle astrometry required to reach the predicted atmospheric limit of 10 μas accuracy at Paranal. Attainment of this performance will allow the reflex motion induced by an orbiting planet on a star located at the distances shown in Figure 8.

This figure shows how the VLTI in its astrometric mode would cope with a planet of a Jupiter, Saturn, Uranus or Gas Giant Core (GGC) of $\sim$10 earth

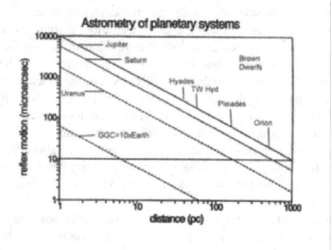

Figure 8. The expected astrometric reflex motion of objects of different masses orbiting a star as a function of the latter's distance. The VLTI is expected to measure reflex motions $> 10 \mu$as.

masses placed at either the Jupiter radius of 5 AU as expected for a solar-like star or closer for a cooler companion. Since the closest star forming regions (SFR) are located between $\sim$50 (TW Hyd) and $\sim$140 pc (Taurus-Auriga) from the Sun, the VLTI would be in the enviable position of easily finding Jupiter-sized planets around stars of any spectral type and, most significantly from the physical point of view, of any age up to the MS. The VLTI then will be able to reach all the way out to Orion to study planets in a wide variety of environments.

The VLTI, by exploiting its dual mode of operation with phase referencing, can, at the same time, probe in considerable detail not only the planet itself through its effect on the motion of the parent star but also the evolution and structure of the proto-planetary/proto-stellar disk from which it eventually must emerge and interact. This affords another unprecedented opportunity to finally understand and pin down the presently uncertain connection between the two phenomena. At the typical distance of the MS early type star IR excess disks, for example, the VLTI IR spatial resolution of a few mas ensures a clear probe of these disks down to the level of a few tenths of AU resolution at a few AU radii where most of the interaction between a possible planet and the edge of the small particle disk is expected to occur.

Other research areas that the VLTI can contribute to significantly and deeper development of many of the topics covered in this document can be found in the proceedings of the ESO workshop *Science with the VLTI*, ESO Astrophysics Symposia Series, (ed.) F. Paresce, Springer-Verlag, Berlin, 1997 and in the document *Scientific Objectives of the VLTI* by F. Paresce that can be found at

http://www.eso.org/projects/vlti/science/VLTIscienceMarch2001.pdf

Acknowledgements

This document could not have been written without the fundamental scientific input of all the members of ESO's Interferometry Science Advisory Committee who contributed in the early years to the establishment of the scientific objectives of the VLTI and to the success of its efforts to get VLTI restarted in 1996 after its near demise in 1993. A particular debt of gratitude is due to Jean-Marie Mariotti whose noble example of dedication to and competence in the field of interferometry inspired many of the ideas in this document and infused the spirit of this new exciting endeavor. The VLTI would not be where it is today without critical contributions from many people too numerous to properly thank. A partial list would include C. Cesarsky, S. Correia, J. Davis, F. Delplancke, F. Derie, E. DiFolco, Duc Thanh Phan, A. Gennai, R. Giacconi, P. Gitton, A. Glindemann, P. Kervella, B. Koehler, S. Menardi, S. Morel, A. Richichi, M. Schoeller, J. Spyromilio, M. Tarenghi, A. Wallander, M. Wittkowski, R. Van Boekel.